The Ontology of Physics for Biology

This book introduces *semantic* representations of multiscale, multidomain physiological systems that link to qualitative reasoning and to quantitative analysis of biophysical processes in health and disease. Two major public health problems, diabetes and hypertension, serve as use-cases to illustrate the depth and rigor of such representations for logical inference and quantitative analysis. Central to this approach is the Ontology of Physics for Biology (OPB) that formally represents the foundations of classical physics and engineering system dynamics that are the basis for our understanding of biomedical entities, processes, and functional relationships.

Furthermore, we introduce OPB-based software for annotating and abstracting available biosimulation models for reuse, recombination, and for archiving of physics-based biomedical knowledge. We have formalized and leveraged physics-based biological knowledge as a working view of physiology and biophysics from three distinct perspectives: (1) biologists and biomedical investigators, (2) biophysicists and bioengineers, and (3) biomedical ontologists and informaticists. We present a logical and intuitive semantics of classical physics as a tool for mediating and translating biophysical knowledge among biomedical domains.

Daniel L. Cook, MD, PhD
John H. Gennari, PhD
Maxwell L. Neal, PhD

The Ontology of Physics for Biology

for Biology

Semantic Modeling of Multiscale,
Multidomain Physiological Systems

Daniel L. Cook, MD, PhD
John H. Gennari, PhD
Maxwell L. Neal, PhD

CRC Press
Taylor & Francis Group
Boca Raton London New York

CRC Press is an imprint of the
Taylor & Francis Group, an **informa** business

Designed cover image: Shutterstock

First edition published 2024
by CRC Press
2385 NW Executive Center Drive, Suite 320, Boca Raton FL 33431

and by CRC Press
4 Park Square, Milton Park, Abingdon, Oxon, OX14 4RN

CRC *Press is an imprint of Taylor & Francis Group, LLC*

© 2024 Taylor & Francis Group, LLC

Library of Congress Cataloging-in-Publication Data
Names: Cook, Daniel L., author. | Gennari, John H., author. | Neal, Maxwell Lewis, author.
Title: The ontology of physics for biology : semantic modeling of
multiscale, multidomain physiological systems / Daniel L. Cook, John H.
Gennari, Maxwell L. Neal.
Description: Boca Raton : CRC Press, 2024. | Includes bibliographical references and index.
Identifiers: LCCN 2023006099 (print) | LCCN 2023006100 (ebook) | ISBN
9781138598058 (hardback) | ISBN 9781032533100 (paperback) | ISBN
9780429469961 (ebook)
Subjects: LCSH: Biophysics–Mathematical models. | Biophysics–Computer
simulation. | Ontologies (Information retrieval)
Classification: LCC QH505 .C635 2024 (print) | LCC QH505 (ebook) | DDC
571.4–dc23/eng/20230802
LC record available at https://lccn.loc.gov/2023006099
LC ebook record available at https://lccn.loc.gov/2023006100

ISBN: 9781138598058 (hbk)
ISBN: 9781032533100 (pbk)
ISBN: 9780429469961 (ebk)

DOI: 10.1201/9780429469961

Typeset in Minion
by codeMantra

Contents

Rationale

Our growing knowledge of biological structure and function spans all spatial scales from genes to organ systems, and physiological domains from membrane ion currents to blood flow. Such a plethora of knowledge challenges our ability to represent and comprehend both normal and pathological functions of the human body. Here, we introduce a novel semantic approach we have implemented as the Ontology of Physics for Biology (OPB) that is based on modern knowledge representational and computational tools to represent and apply the rules and theories of classical physics to the solution of physiological problems in bioscience and biomedicine.

Preface – What We Are Talking About, and Why

We describe a semantic architecture and model representation tools as developed over the past 10 years by the Semantics of Biological Processes group at the University of Washington. Our work incorporates lessons and works from our participation in the DARPA Virtual Soldier project, the EU Virtual Physiological Human project, and the NIH Virtual Physiological Rat project. Each of these projects demanded an accounting of the widest range of physiological systems and biophysical processes spanning all structural scales and physiological domains. Here, we describe (1) our novel semantic knowledge representation methods upon which we have (2) built the Ontology of Physics for Biology and (3) implemented and deployed a suite of computational tools for building a computable semantic Physiome.

This book is a primer for undergraduate, graduate, and professional students that surveys the basics of three disciplines – bioinformatics, biomedical ontology, and biophysical modeling. It proposes and develops novel semantic methods for the representation and analysis of physics-based multiscale, multidomain physiological systems in health and disease.

We aim to represent three typically orthogonal approaches to representing, understanding, and computing the scientific basis of physiological and biophysical entities and processes. We are inviting workers in each of the three concurrent fields of study to better understand, appreciate, and collaborate across traditional knowledge domains. We are concerned with three traditionally "siloed" disciplines that operate with little crosstalk, understanding, or appreciation for the approaches and contributions of the other disciplines.

1. **Biomedical clinicians and researchers** who strive to describe, understand, and, hopefully, relieve suffering and cure patients of diseases and pathophysiologies that are important public health problems. Their approach is largely observational, intuitive, and descriptive.

2. **Bioengineers and biophysicists** who bring powerful computational methods of engineering system dynamics and classical physics for representing, analyzing, and explaining normal and pathological physiological systems. Their approach is analytical with a reliance on largely quantitative methods.

3. **Bioinformaticians** whose prescript is to organize, express, archive, and provide access to a rapidly expanding compendium of biomedical and biophysical knowledge, and to develop and deploy bioinformatical computational tools for knowledge representation and analysis.

Glossary

Here we briefly define basic terms and concepts that apply to the work that we discuss.

"**Biology**" – We define "biology" as "Biology is the natural science that involves the study of life and living organisms, including their physical structure, chemical composition, function, development, and evolution." Within this broad mandate, we focus on biological organisms and systems that are of interest to the biomedical community either as clinical subjects or as research models.

"**Physiological system**" In the broadest sense, a "physiological system" includes one or more material and immaterial objects of biological origin – whose existence and composition depends on gene expression – that interact by exchange of momentum, material, thermodynamic energy, and/ or information. We interpret this definition broadly to include ancillary technology by which physiological systems are observed and manipulated.

"**Physics**" means "classical" physics as discovered and articulated, primarily in the 19th century, by Newton, Ohm, Hooke, and others, and as taught in college physics classes. We focus on biomedical systems dynamics, the application of physical principles for the analysis of dynamical systems as developed and applied to the engineering sciences. We exclude particle, radiation, and relativistic physics that find only limited applications for physiological modeling. Furthermore, we limit ourselves to *discrete* systems that are usefully analyzed by temporal differentials and temporal differential equations.

"**System dynamics**" is a branch of engineering systems analysis that has broad applications in the engineering and biophysical sciences. It is primarily concerned with analyzing the *stock* and *flow* of stuff such as money, manufactured items, blood in the circulatory system, or cellular metabolites.

"**Modeling**" is the abstraction and computational representation of how physical things are structured and how they participate in processes.

In the biophysical and physiological sciences, it has been a foundational tool of knowledge representation, simulation, and analysis since the 1950s.

"*Semantics*" is the assignment of *meaning* to words, signs, things, and symbols that is the basis for mapping between various biomedical knowledge resources in order to correlate the knowledge content of the different resources. For our applications, we seek to recognize the semantic equivalence of, for example: (1) a *verbal statement* that "increasing blood sugar levels increases insulin secretion rate", (2) a *data plot* or *statistical model* that correlates insulin secretion rate to a glucose stimulus, (3) a *cause–effect diagram* that represents glucose to insulin secretion, or (4) the mathematical dependence of a model variable for insulin secretion rate, "IR" to a glucose concentration variable, "G".

"*Ontology*" is a branch of philosophy and computer science for formally naming and defining types of entities and processes, assigning properties to each, and establishing relationships among them. Computational ontologies have emerged as the preeminent tool for knowledge representation in many biomedical domains that consist of a taxonomy of classes with logical relations between them.

WHO WE ARE

Dan, John, and Max have been collaborating for over 15 years in the overlapping areas of physiology and biophysics, bioinformatics, ontology, knowledge sharing, and systems modeling. Their scientific collaboration began as a collaboration on the DARPA-sponsored Virtual Soldier Project (VSP), led by Professors Cornelius Rosse and James Bassingthwaighte at the University of Washington. This experience established their main aims of developing an ontological representation of physics, integrating it into the burgeoning field of biomedical ontology, and creating software to support semantic causal analysis of physiological systems.

Dan is an Emeritus Professor of Physiology and Biophysics at the University of Washington, Seattle. He graduated (BSME, 1967) from the University of Michigan Mechanical Engineering and worked at the Boeing Airplane Company, first, to manufacture the first 747 airliner, and then to analyze the structural dynamic of the (unbuilt) Boeing supersonic transport (SST). Taking an interest in the emerging field of bioengineering, he earned a master's degree in mechanical engineering (UW, MSME, 1971) modeling the cellular dynamics of insulin secretion. He then entered the UW's Medical Scientist Training Program (MSTP, 1971) to earn MD and PhD degrees. He has published seminal laboratory and modeling studies of the electrophysiology of insulin secretion and of auditory sound localization. He is retired and lives in Seattle with his wife.

John is a Professor and Graduate Program Director for Biomedical & Health Informatics (BHI) at the University of Washington. His background is in computer science and artificial intelligence, and was introduced to the field of biomedical informatics in the early 1990s at Stanford University. There, he developed an interest in knowledge representation as applied to biomedical applications and collaborated with early developers of ontologies. After joining the University of Washington in 2002, he began his collaboration with Max, Dan, and Cornelius Rosse around models of anatomy and physiology. In addition to teaching and leadership roles in BHI, John continues to be active in research, furthering efforts in standards development and reproducibility. John enjoys Seattle and the Pacific Northwest with his family.

Max is a Senior Scientist at Seattle Children's Research Institute. Since his first exposure to dynamic physiological modeling while working on DARPA's VSP, Max's work has focused on applying computational methods to understand various biological systems as well as the development of standards and tools that facilitate systems-level biological modeling. Meeting and collaborating with John and Dan during the VSP established his long-standing interest in semantics-based representations of biosimulation models, which he studied for his PhD at the University of Washington. Since then, he has led the adoption of community-ratified metadata standards for biosimulation and models as well as the development of software for semantics-based biological modeling. He lives in the Seattle area with his wife and son.

CHAPTER-BY-CHAPTER OVERVIEW
Orientation – What to Expect

Our goals are pragmatic. We aim to build tools and resources that are useful to biomedical investigators, educators, and students who can benefit from a foundational theory of classical physics for research, computations, and practice in biomedicine. We recognize the challenges of the incredible complexity of multiscale (molecules to organisms) and multidomain (fluids, chemical kinetic, electrophysiology, etc.) systems whose structure and function need to be understood and represented. To maximize utility, generality, and capability of our tools, we address multiscale/multidomain problems that are of intense biomedical interest (Chapter 1) using the expressive power of modern bioinformatical tools (Chapters 2–4).

We adopt the analytical power and rigor of classical physics as implemented by bioengineering systems dynamics (Chapter 5) whose rules we formalize as the Ontology of Physics for Biology (OPB; Chapter 6)..

Chapter 1 – Biomedical Challenges, Solutions

This chapter is a "primer" that describes two biomedical and public health challenges to understanding, representing, and managing two urgent public health concerns – systemic hypertension and diabetes mellitus. These problems are *multiscale*, involving physical structure from atoms to whole organisms, and *multidomain* involving biochemistry, cell biology, and systems physiology. Studying, articulating, and solving such problems require a broad range of biomedical information, knowledge, and analytical tools that are, unfortunately, sequestered in various knowledge and technological "silos".

Chapter 2 – Biomedical Information and Data Resources

We survey the extensive range of on-line and in-print resources of biomedical knowledge including databases, terminologies, and ontologies. We describe specific resources – databases, terminologies, and knowledge bases – that are most germane for representing and describing physiological systems in health and disease.

Chapter 3 – Biomedical Ontologies

Most students and investigators are familiar with the structure and function of databases and terminologies, but are less familiar with the formal structure and uses of biomedical *ontologies*. Thus, Chapter 3 is, then, a primer on the development, structure, and use of ontologies to organize and standardize biomedical knowledge. Beyond terminologies, ontologies represent not only things, such as anatomical parts, but also the relationships between those things.

Chapter 4 – Biophysical Modeling

This chapter reviews the quantitative, physics-based analytical methods of physiology, biophysics, and bioengineering to analyze and simulate complex functional hypotheses of normal physiology and pathophysiology of disease. This chapter emphasizes the broad range of methods used to represent and compute analytical models.

Chapter 5 – System Dynamic Modeling

This chapter extends Chapter 4 by describing the basics of *system dynamics* as a subdiscipline of classical physics developed by the engineering sciences with applications for the practical simulations of physics-based, stock-and-flow models of multiscale/multidomain biophysical systems.

Chapter 6 – Ontology of Physics for Biology

This chapter introduces the Ontology of Physics for Biology (OPB) and describes its representational principles as applied to the rules and theories of classical physics. Although its development has been motivated by biomedical use-cases, the foundations of the OPB apply to and can be extended to encompass other domains. OPB ascribes to and leverages available biomedical methods and standards as needed for the semantic annotation and analysis of available biophysical data, analytical, and modeling resources.

Chapter 7 – OPB-Based Semantic Modeling

We introduce our SemGen software for annotating, analyzing, archiving, and reusing biophysical models to support collaboration of international teams of biomedical researchers, bioengineers, and bioinformaticians. We also discuss the contributions of our semantic modeling methods to efforts of international consortia for documenting, archiving, and sharing biophysical knowledge for both clinical and basic biomedical research.

Chapter 8 – Summary and Conclusions

Motivated by the need to represent and analyze the pathophysiology of two clinically important diseases of multiscale, multidomain physiological systems, we recapitulate the need for a semantic approach to physiological modeling for unifying available quantitative knowledge of available mathematical models with available qualitative knowledge of existing biomedical resources.

CHAPTER 1

Biomedical Challenges

S INCE THE ANCIENT TIMES of Galen, biomedicine has been the field of scientific and clinical study that aims to discover, understand, and express knowledge about biological organisms in health and disease. In modern times, the challenge has been to reach for these aims in the face of extraordinarily large and complex aggregations of data about molecules, cells, organs, and organ systems. These data are complex in part because they are often at very different scales, both in time and in space: how quickly a physiological process occurs, and how small or large are the biological participants in that process. Furthermore, we have learned that many common pathologies (e.g., diabetes or hypertension) cannot be effectively studied or understood in isolation. Rather, these diseases affect and are affected by many biological organs, systems, and processes. In part because of this complexity, recent computational advances in our ability to store, access, and compute with this data have not always led to corresponding advances in our understanding of the underlying mechanisms of disease and physiology.

In this book, we describe mathematical and computer-based approaches to modeling physiology and biophysics. We believe that to accelerate research and to achieve new insights about the mechanisms of disease, we must develop the ability to integrate computer-based modeling of physiology across scales and domains. To this end, we present the *Ontology of Physics for Biology*, a standard for representing the underlying semantics and physical constraints that all biological systems share. It is our premise that the use of this ontology will enable improved sharing, integration, and understanding of biological models. In spite of the complexity and scale of modern biological data, we believe that modern analytical tools paired with common semantics will enable improved computational and mechanistic models. In turn, these models should lead to new insights, accelerate our understanding of pathophysiology, and, perhaps, be a guide to better diagnosis and treatment.

REAL-WORLD USE-CASES – THE NEED FOR UNDERSTANDING

To motivate this book and set its aims, we describe the challenges of describing, understanding, and managing two serious public health concerns: *essential hypertension* that is characterized by high blood pressure (HBP) and its pathological consequences, and *adult-onset, or type 2 diabetes mellitus* (DM, type 2 DM). We focus on these highly prevalent diseases that, although mostly treatable, continue to be responsible for an inordinate amount of public health concern and expense as well as dire impacts on the health of individual patients. We focus on these diseases as use-cases because both are (1) *multiscale* in that each disease spans broad spatial scales (microns to meters) and temporal scales (nanoseconds to years), (2) *multidomain* in that each involves processes across multiple biophysical domains including chemical kinetics, fluid transport, and solid mechanics, and (3) each disease is understood to result from failure of physiological control of blood glucose level in diabetes and of blood pressure in hypertension.

Major features of the detection and physiological regulation of blood pressure and blood glucose are shown in Figure 1.1. The key clinical and physiological properties of blood glucose concentration and blood pressure are assessed using routine clinical measures such as (1) arterial blood pressures by pressure cuff (bottom right), (2) tissue blood using "glucometers" (bottom, middle), and (3) venous blood glucose by venipuncture and lab analysis (bottom, left). Each of these measures a property of blood that is pumped by the heart (top, middle) through the arterial circulatory system (right hand, descending arrow), through networks of tissue arterioles, capillaries, and venules (bottom, middle network) to be returned via the venous system (left hand, ascending arrow) to the lungs (not shown) and the heart.

Establishing and regulating both blood fluid pressure and blood glucose concentrations depend on many tissues but primarily on tissues and organs such as (left-to-right; middle panel): (1) kidneys with their attached adrenal glands, (2) skeletal muscle, (3) liver, (4) body fat deposits and fat

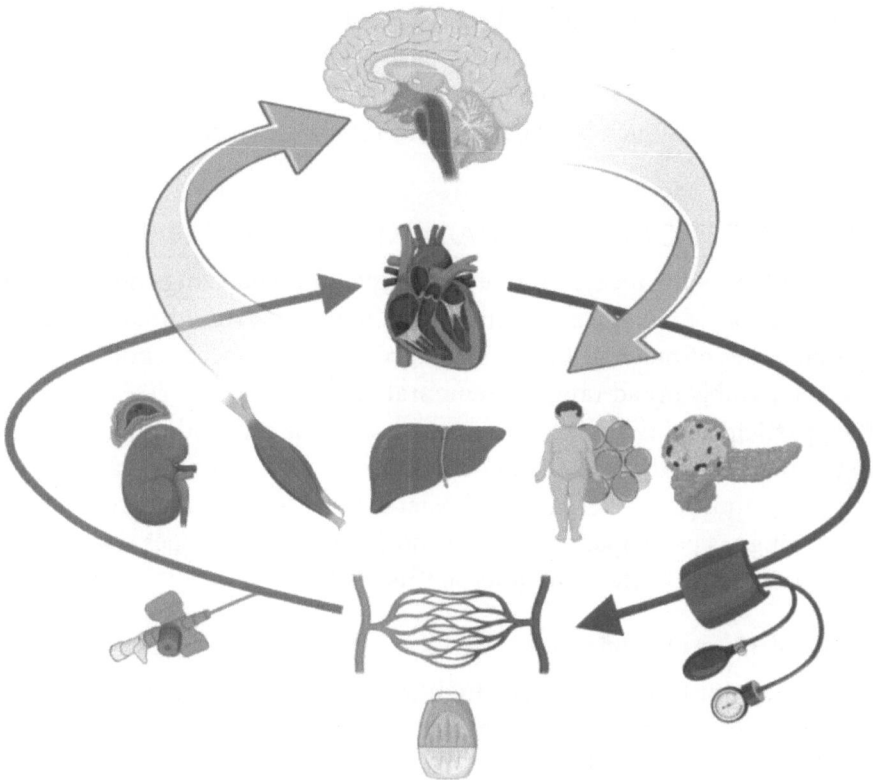

FIGURE 1.1 Cartoon of major anatomical players, physiological pathways, and diagnostic instruments that sense and control blood pressures and glucose levels (see the text).

cells, and (5) pancreas and its embedded, hormone-secreting pancreatic islets. The physiology of each of these organs depends (right-side, descending arrow) on either direct neural control via autonomic nerves controlled by (top) brainstem centers and/or by hormones released by the brain's pituitary gland – the so-called neurohormonal axis.

These responses depend primarily on the reading of the hormonal and autonomic nerve inputs by the hypothalamus for monitoring the internal physiological state of the body. This broad set of functions, represented by the broad ascending and descending arrows in Figure 1.1, are now termed "interoception" (Zaman et al., 2022) and consist of networks of autonomic nerves that project to the brainstem along with blood-borne hormonal signals from various organs.

Physicians and biomedical scientists have been productively studying these complex processes both as a basic scientific enquiry and to articulate approaches for intervention and cure. To store, access, and use this knowledge, biomedical researchers have partnered with their computer and informatics coworkers to create, deploy, and distribute a variety of bioinformatics tools such as databases, terminologies, and, most recently, ontologies.

SCOPE OF THE MULTIDOMAIN, MULTISCALE CHALLENGE

The challenges for measuring, representing, and analyzing the structure and function of biological systems stem from the extraordinary range of spatial scales of material body parts – from atoms to whole bodies – and the comparably broad range of temporal scales of the processes in which they participate. Figure 1.2 presents a view of the physical things according to their sizes from atoms to bodies, how these things are classified by type (e.g., molecule or organ), and what kind of biological processes have such things as participants. These are some of the basic issues that we will address in detail in the remainder of this book.

Types of Things (Continuants)

Figure 1.2 illustrates, at a very high level, the multiscale, multidomain challenges of studying, modeling, and understanding normal and pathological processes in health and disease. We start by noting the breadth of spatial scale ($\approx 10^{13}$) of biological *things* that range from atoms to bodies of

Physical things – spatial scale in meters

atom	molecule	cell-part	cell	organ	body

10^{-12} 10^{-9} 10^{-6} 10^{-3} 1   10

Classes of things – count

> 100 atom classes	>> 100,000 molecule classes	> 400 cell-part classes	> 600 cell classes	63 organ classes	12 systems classes	2 body classes

Kinds of processes — time scale in seconds

molecular motion	diffusion neural signaling	cell contraction secretion	heart beat skeletal motion	endocrine regulation	disease progression	organism lifetime

10^{-6} 10^{-3} 1 10^3 10^6

FIGURE 1.2 "Multiscale, multidomain" physiology and biophysics: (top) kinds of physical things over a broad range of spatial scales (>1,012), (middle) the cardinality of kinds of "things", and the time scales of the physical processes in which those things participate (bottom).

organisms that are of concern to biomedical and physiological investigators (top panel). Furthermore, within each class of thing, there are many subclasses that range from 2 body types (e.g., male, female) to well over 100,000 different molecule classes (e.g., from ions to proteins and genes) that must be identified, characterized, and named.

Types of Processes (Occurrents)

Even more daunting is the challenge of naming, defining, and counting multidomain, multiscale *processes*. For example, one could define a simple model of blood pressure regulation as some algebraic or differential equation relating the value of blood pressure to heart rate, and then elaborate the model by including heart rate and ventricular contractility, and then by a more detailed model of myocardial muscle mechanics. The point is that process definitions are a daunting, open-ended combinatorial problem because each participant in a process depends, in turn, on the participants in one or more other processes.

Physiological Domains

Figures 1.1 and 1.2 illustrate the scope of bodily systems and processes that participate in the normal regulation of blood pressure and blood sugar levels. The processes span several *physiological domains* pertinent to normal function and the pathophysiology of disease. Multidomain problems such as these are challenging simply because they span traditional academic and disciplinary domains that are distinct knowledge "silos" each with its own preferred vocabulary and jargon, data storage and analytical methods, and approaches to problem solving.

Physical Property Measures and Analysis

Our knowledge of physiological things, processes, and domains is based on empirical observation, measurement, and analysis of the physical properties of things and the processes in which they participate. In this book, we are concerned exclusively with the use of physics-based property values and the classical physical laws by which such property values depend upon one another. Chapter 4 reviews and discusses the methods by which physiologists, biophysicists, and bioengineers quantify the values of physical properties. Chapter 5 describes how they have used the theories and methods of engineering systems dynamics to model and test physics-based models of multiscale, multidomain physiology in health and disease.

Understanding Systems, Predicting Outcomes

In this book, we describe the modern biomedical informatics tools that formally describe, reason about, and compute on the functional content of multiscale, multidomain physiological systems in health and disease. These resources offer formalized resources for developing physics-based explanations of these complex phenomena that are available for teaching students and for directing research and explaining results. As we will describe in Chapters 2 and 3, some of these challenges can be addressed using the available tools of biomedical informatics including biomedical databases, taxonomies, and ontologies.

We present these scenarios as preludes for the main aim of this book, which is to describe modern computational tools of bioinformatics and modeling are used to (1) represent the things – molecules, cells, organs, systems, etc. – that comprise the physiological systems, (2) how the behaviors and functions of such systems as they are recorded, modeled, and analyzed, and (3) we present novel bioinformatics tools to support physics-based computational systems modeling and functional reasoning about complex biological systems in health and disease.

The Challenges

Domain and knowledge boundaries present real impediments to collaborative solutions, particularly for multidomain problems such as diabetes and hypertension because of field-to-field differences in conventions for, among others: (1) anatomical naming, boundaries, and terminologies, (2) units and scaling for measurement data, (3) theoretical abstractions for parts and processes, and (4) methods for data and information representation, storage, and retrieval. Whereas many of these knowledge representation differences can be sufficiently resolved for informal communication, computational and database problems are being resolved using bioinformatical tools to be reviewed in Chapter 2 and biomedical ontologies to be reviewed in Chapter 3.

USE CASE 1: HYPERTENSION AS FAILED BLOOD PRESSURE CONTROL

The cardinal sign of essential hypertension is elevated arterial blood pressure that, if untreated, can accelerate arterial atherosclerotic plaque formation, stroke and heart attack, and death. All who have visited a medical clinic have had their blood pressure measured as, say, "120 over 70", which characterizes the peak (systolic) and trough (diastolic) pressure values of arterial blood flow and pressure pulses. Hypertension is diagnosed by abnormally high systolic and diastolic blood pressures (i.e., greater than 140/90 mmHg). Although some causes of hypertension can be identified, most commonly, hypertension is attributed to "essential hypertension" that can be diagnosed and treated, but cannot be cured. Thus, chronic hypertension remains a major public health problem that accounts for accelerated atherosclerotic damage to arterial walls throughout the body and brain. Such damage is particularly of concern when it occurs in the arterial vessels of the brain and in the coronary vessels that supply blood to heart muscle tissue.

The circulatory system distributes blood to all body tissues with which it exchanges respiratory gasses (oxygen, carbon dioxide), nutrients, and blood-borne hormones among all tissues. These critical transport and exchange functions require a *regulated* balance of the heart's blood-pumping effort with the blood flow resistance of the arterial vascular systems. One could expect that high arterial pressure is due to excess heart pump effort, or excess blood vessel constriction, or both. However, this analysis only begins the search for an answer to the problem given the intricacy of redundant and parallel networks of blood flow and pressure regulatory pathways. In the following, we will expand our description of cardiovascular physiology with respect to hypertension and then discuss analytical and informatics tools that are used to represent and analyze the physiology and pathophysiology of hypertension.

Arterial Hypertension – Anatomy and Physiology

Deviation of these clinical measures has driven extensive efforts to understand how blood pressure is regulated and to identify what changes occur that result in clinical hypertension. Some changes occur over a lifetime such as the progressive build-up of atherosclerotic plaques that reduce blood flow and can disrupt how the kidneys properly sense and respond to changes of blood pressures. Other responses occur on a time scale of seconds such as during a "Valsalva maneuver" that pressurizes the thoracic cavity, increases blood pressures, and triggers a series of regulatory events beginning with the decreased heart rate, relaxed arterial walls, and a compensatory fall in blood pressure.

The cardiovascular system, as diagrammed in Figure 1.3, consists of the heart that pumps blood through arteries and back through veins to transport nutrients, respiratory gasses, and waste products to and from

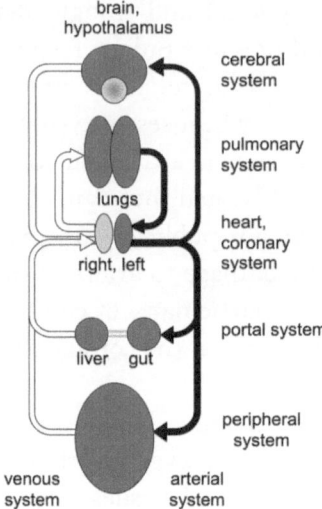

FIGURE 1.3 The cardiovascular system consists of the arterial system (diagram right side) that carries oxygenated blood pumped by the left ventricle to all tissues, and the venous system (diagram left side) that carries oxygen-depleted blood back to the right side of the heart to be pumped through the lungs and back to the left side of the heart.

organs and tissues throughout the body. It comprises two serially connected blood flow paths: the pulmonary flow path carries blood from the "right" side of the heart, through the lungs and back to the "left" side of the heart (left vs. right are from the perspective of the patient; not of an observer).

The left side of the heart pumps blood through the "systemic" arterial system to all body tissues and then back to the right side of the heart through the venous system. The multi-chambered, muscular heart pressurizes the blood in the left ventricular chamber to force blood to flow through the "aortic" valve, into the aorta, and out to the rest of the body. The heart ventricles contract periodically to pump blood through the outflow pathway on the other side of the outflow valve – the aortic valve on the left side; pulmonary valve on the right side.

The Cardiac Cycle

The term "cardiac cycle" refers to the physiological events that occur and recur during a heart beat in which blood is simultaneously drawn into both left and right atria and then expelled by contraction of the muscular walls of each atrium. Blood is expelled through one-way atrioventricular valves into the ventricles and then expelled from the ventricles by the contraction of the myocardial muscles in the ventricular wall. The periodic contractions result in periodic pulses of blood pressure that are readily detectable, and familiar, as heart "beats" that reflect the periodic "systolic" pressurization of blood followed by the "diastolic" nadir as blood flows on through the vascular system (Figure 1.4).

> *Pathology.* The ordered myocardial contractions of the heart chambers can be severely disrupted by changes in the balance of membrane ionic currents that excite myocardial muscle cells to contract. One serious failure is "ventricular fibrillation" due to uncoordinated triggering of ion channel activity that fails to trigger coordinated ventricular muscle contraction.

Myocardial Excitation

This intricate physiological dance must be properly performed to maintain systemic blood flow and pressure. Both the strength and frequency of contractions are regulated on a beat-to-beat by neurohormones and nerve input from the brain stem. Simultaneously, a wave of electrical atrial muscle cell membranes spreads from the atrium's sinus node via a bundle of specialized muscle fibers that trigger the coordinated contraction of the left ventricle myocardium to pressurize blood in the ventricular cavity.

> *Pathology.* Ventricular fibrillation, far more serious than atrial fibrillation, is the uncoordinated electrical contractile activity of left ventricular myocardial cells that defeats the coordinated ventricular muscle contraction to pump blood through the systemic circulation often with fatal consequences.

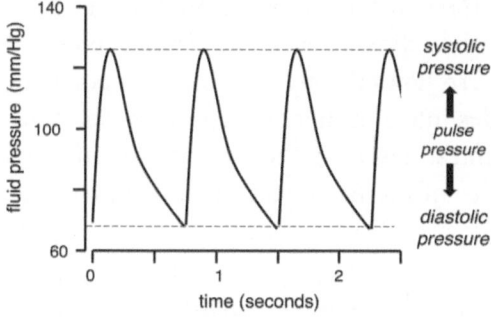

FIGURE 1.4 Left ventricular contraction (systole) drives blood through the aortic valve to transiently pressurize blood in the aorta and propels blood flow onto the arterial system.

Ventricular Contraction, Aortic Valve

During the left ventricular systolic stroke, ventricular blood pressure exceeds that of aortic blood, so that the aortic flap valve opens, and blood surges into and pressurizes the blood in the proximal aorta. The aortic valve is a one-way valve that prevents backflow from the aorta into the ventricle, which produces the first heart sound of the cardiac cycle.

> *Pathology.* Aortic valve stenosis is a consequence of wear and tear on aortic valve leaflets that is often due to prevailing hypertension and to bacterial infection of the leaflets themselves. Aortic stenosis is a serious condition that limits blood flow to the brain and body leading to muscle weakness and brain ischemia and stroke.

Blood Flow into and through the Proximal Aorta

The injection of a pulse of blood into the aorta expands its elastic walls with an attendant increase in blood pressure which drives blood flow into and along.

> *Pathology.* High aortic blood pressure, flow rate, and subsequent turbulence exert wear and tear on the vessel wall that can lead to an *aortic aneurysm* that is a pathological ballooning and thinning out of the vessel wall that can harbor and throw dangerous blood clots or, worse yet, burst and cause sudden death by exsanguination.

Blood Flow through the Arteries, Arterioles, and Capillaries

The cardiovascular system circulates blood throughout the body to distribute essential fluids, oxygen, nutrients, and hormones to all cells in the body while delivering their metabolic waste products to the liver and kidneys and delivering carbon dioxide back to the lungs. Blood, pressurized by the left side of the heart, flows through the one-way aortic valve into the aorta which then divides into a number of arterial vessels that supply the digestive tract (liver, intestines, etc.), the kidneys (left and right), as well as the bones and muscles of the head/neck, abdomen, trunk, and legs. Arterial blood delivers oxygen and nutrients throughout the vascular tree as it bifurcates many times until reaching arterioles into the capillaries in tissues and organs before being collected by the network of venules and veins to return via the vena cava to the right side of the heart. Blood returning to the heart is then pumped through to the lungs where carbon dioxide, a waste product of fuel metabolism, is exchanged for oxygen in freshly inspired air.

Pulsatile Blood Pressure and Flow through the Vascular System

Arterial blood pressure is of critical clinical importance. Too little pressure (hypotension) and blood flow rates can be too low to properly perfuse tissues, particularly the brain. Too much pressure (hypertension) is taxing for the myocardium and the arterial walls that can increase atherosclerosis and the incidence of cerebral vascular strokes. As blood flows from the aorta (left) into the capillaries (middle) and into the veins (right), there is a dramatic fall in mean blood pressure and pulse pressure due to the viscous drag of blood moving along vessel walls (Figure 1.5).

As vessels bifurcate, their diameters and cross-sectional areas decrease (Figure 1.6). However, they become so much more numerous that the sum of their cross-sectional and surface areas increases dramatically and flow velocity decreases accordingly to allow sufficient time and surface area to assure needed fluid and material transport into the surrounding extracellular tissues spaces.

Arterial Pulsatile Flow; Vascular Stress

As blood flows through the branching segments of the arterial tree, blood pressure and flow pulsatility are damped down as each segment absorbs and slows the flow to the next segment. Blood flows into the smallest arterial vessels, the arterioles, that actively constrict and relax as controlled by neural signals from brain stem signals and by factors that depend on local tissue needs.

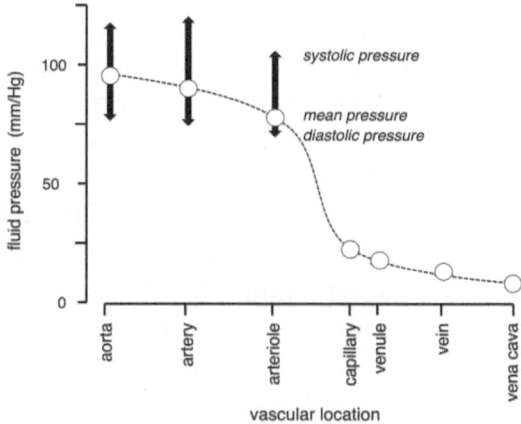

FIGURE 1.5   Progressive decrease of mean blood pressure and pulse pressure as blood flows from the aorta into arteries through arterioles and capillaries, and into the vena cava.

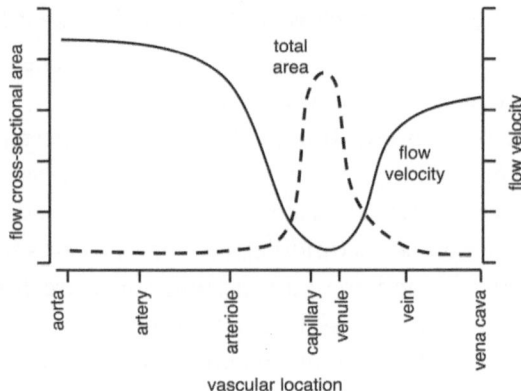

FIGURE 1.6   Schematic diagram of the vascular system and its corresponding blood pressures and blood flow rates.

The periodic contraction of ventricular muscle and subsequent outflow of blood causes a periodic peak in aortic blood pressure that decays as blood flows out through the aorta and into downstream parts of the arterial system. The same pulsatility occurs in the pulmonary artery but at a far lower blood pressure. The most accessible indicator of cardiovascular function is the *pulse* observed, for example, over arteries in the wrist, neck, groin, or ankle that provides immediate signs that the heart is beating and that there is an intact vascular flow path from the heart to other body parts. Second, the heart rate, in beats per minute (bpm, normally about 72 bpm), is an index of cardiovascular health. A fast heartbeat (tachycardia) is a normal response to physical exertion or may indicate a pacing defect in the heart itself. A slow heartbeat (bradycardia) may indicate a disruption of normal pacemaker function within the heart itself.

Arterioles Regulate Blood Pressure and Flow Rate
As blood flows through the branching segments of the arterial tree, both blood flow and blood pressure pulses are diminished as each segment periodically fills and empties at rates proportional to the blood pressure differential across the vessel and inverse to the blood's viscosity. After many arterial bifurcations into smaller and shorter arterial branches, blood flows into arterioles that are the smallest of arterial vessels. Arterioles have muscular walls whose contraction and relaxation control blood flow into capillaries where fluids, respiratory gasses (oxygen, carbon dioxide), and nutrients are exchanged with the extracellular fluid bathing cells and tissues. Arteriole caliber is controlled by neural inputs from the sympathetic

and parasympathetic nervous systems that originate from blood pressure regulatory centers in the brainstem.

Systemic Control of Blood Pressure and Flow

All body tissues require sufficient blood flow to deliver nutrients and oxygen and to remove metabolic waste products and carbon dioxide. Failures of any aspect can have functional as well as fatal consequences. In principle, healthy cardiovascular function depends on maintaining key physiological processes that are subject to interlocking physiological controls that are based on system dynamical principles such as

1. heart *pump flow rate* must sustain blood *pressure* to meet overall body demands,

2. blood *volume* must be maintained for effective heart pumping and total blood flow, and

3. local vascular flow resistances for minute-to-minute tissue *flow rate* demands.

These physiological demands are recognized and modeled in terms of observable physical properties of (pressures, flow rates, etc.) and are inputs to a distributed, multiscale feedback system of neurohormonal and fluid dynamical processes pathways at all anatomical levels. For example, the neural and hormonal pathways that regulate cardiovascular function include: *baroreceptor* (pressure sensors) in the walls of aorta and carotid arteries sense blood pressure and signal neurons of the brainstem nucleus tractus solitarius (NTS). The NTS then serves as a feedback controller of heart pump rate and contractile strength mediated by the combined effects of vagus nerve and sympathetic nerve efferents.

The key regulated variable for controlling blood pressure can be continuously and directly measured using intravascular electronic pressure probes to distinguish and measure peak systolic and nadir diastolic pressures. Whatever the measurement method, the relationship of systolic to diastolic pressures is diagnostic for (1) normal is about 120/70 mmHg, (2) "systolic hypertension" may indicate either a "hyperkinetic" heart with a too vigorous contraction or a normal contraction pushing against a too rigid, "hardened" arterial system, or (3) a "hypokinetic" heart due to restricted outflow through diseased aortic valve, a diseased and weakened heart muscle, or a dilated ventricle with a thinned and weakened ventricular wall.

The cardiovascular system must sustain blood pressure sufficient to assure blood flow to body tissues to meet the physiological demands that range from minimal during sleep, to elevated during physical exertion, to the mortal demands of trauma. One basic operating principle is that of forcing water flow through a drinking straw – the flow rate through a vessel segment depends directly on the pressure drop across the segment and inversely on the fluid flow resistance. Heart ventricles must exert sufficient contractile force to pressurize blood to flow out of the ventricles against the pressures in the blood in the pulmonary and systemic arteries. Blood flow through the arterial trees is regionally distributed to tissues according to the contractile activity of local arteriolar sphincters so that needy tissues get a larger portion of total blood flow.

USE CASE 2: DIABETES MELLITUS AS FAILED BLOOD SUGAR CONTROL

"Diabetes mellitus" is a disease that may be chronic and progressive or acutely fatal. It was first recognized by the ancients in Egypt, India, and Greece. It is characterized by high volumes of urine flow ("diabetes" meaning "flow through") and the sweetness of urine ("mellitus" means "sweet") due to loss of glucose and fluid via the urine. The singular goal in diabetes research and clinical practice is to understand and treat the physiological basis of poor blood glucose control.

Diabetes appears in two forms. Juvenile-onset diabetes (type 1 diabetes) is due to an utter failure of the pancreas to synthesize and release insulin leading to severe and intractable hyperglycemia that is treatable only with injections of the hormone insulin. Adult-onset diabetes (type 2 diabetes) is highly prevalent as a major public health problem leading to serious health consequences such as nerve damage, visual problems, accelerated atherosclerosis, stroke and heart attacks, and premature death. Although it does run in families, its major risk factors are overeating, inactivity, and obesity; hence, there has been a steady increase in the prevalence of type 2 diabetes in the past few decades following increased societal wealth and availability of food.

Clinical Test for Elevated Blood Sugar

Just as the Valsalva maneuver can reveal important dynamical aspects of cardiovascular regulation, so too can various clinical tests for plasma glucose regulation and diabetes. One example is the oral glucose tolerance test (OGTT) in which a standard amount of glucose is ingested by the subject

(Figure 1.7). Then, over the course of *hours*, intravenous blood samples are drawn and analyzed for glucose levels to characterize how well glucose is regulated to stay in a normal range. For normal subjects, the "resting" glucose level will be in the 80–100 mg/dL range and will not exceed 150 mg/dL as pancreatic insulin secretion is sufficient to facilitate glucose uptake by fat and muscle tissues. However, with increasing impairment, too little insulin is secreted and its effects on fat and muscle diminished (insulin resistance) so that excess levels of glucose are seen.

Diabetes is a common metabolic disorder and a public health problem characterized by high blood sugar (glucose) levels due to failure of the normal feedback control by the feedback control of the hormone insulin. Failed insulin feedback control leads to hyperglycemia, the accumulation of excess glucose in the blood. In the short term, this can lead to the often fatal diabetic ketoacidosis. In the long term, it leads to accelerated atherosclerosis and increased rates of heart attack, cerebral stroke, kidney failure, nerve damage, and other systemic pathologies.

Multiscale, Multidomain Pathophysiology of Diabetes

Such pervasive disruptions stem from the central role of glucose and its metabolism in the overall metabolic economy of the body. All cells store and metabolize glucose as an energy-rich carbohydrate that is absorbed from meals, transported in the blood, and stored and released from fat cells according to the cell's need for metabolic energy. The concentration of glucose dissolved in blood must be maintained within narrow limits to

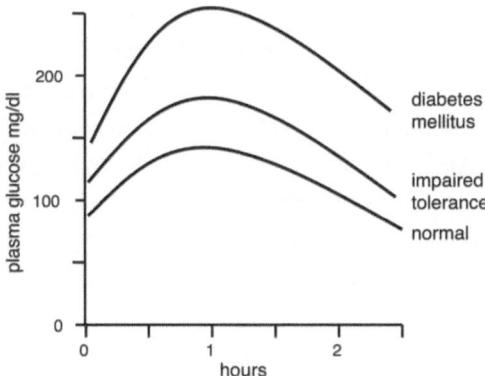

FIGURE 1.7 Illustration of how the basic OGTT distinguishes plasma glucose profiles in normal subjects and those with impaired glucose tolerance or diabetes mellitus.

sustain; too little, hypoglycemia, and cells that require glucose for energy will starve; too much, hyperglycemia, and the excess is transformed into excess fat tissue (obesity) and atherosclerosis.

The pathophysiology of diabetes is a multiscale, multidomain challenge as it involves participants that span anatomical scales from organs interacting via the blood circulation system, organs that exchange nutrients and waste products, cells that communicate signals and biochemicals, cellular metabolic systems, cell membrane transport systems for exchanging molecules and chemical ions between cell interiors and extracellular spaces.

The control of body weight, appetite, and blood sugar levels each depends on a set of interacting neural and hormonal control circuits that rely, in part, on insulin concentration in the blood as a primary feedback regulatory signal. Insulin is produced and secreted by endocrine pancreatic tissue that is analogous to a thermostat for regulating blood glucose levels. As glucose is absorbed from a meal, the increasing blood glucose levels trigger the secretion of the peptide hormone insulin into the bloodstream. Insulin is carried by the bloodstream as a signal for other tissues (notably muscle, fat, etc.) to absorb glucose and return its level back to normal. This function is one example of "homeostasis" that is the stable maintenance of the internal and external environments of cells and organs. In the short term, too much insulin drives blood glucose levels down enough to cause coma and death, too little insulin leads to "ketoacidosis" which is also fatal.

Pancreatic Hormones as Blood Glucose Controllers

The pancreatic beta cells exist in the pancreatic islets of Langerhans scattered throughout the pancreas. Beta cells have evolved to monitor and integrate a host of metabolic, hormonal, and neural input signals and to secrete the hormone insulin at a rate that varies as needed to regulate the level of blood glucose. Failure to do so leads to hypoglycemia (low blood sugar) and subsequent loss of consciousness or to hyperglycemia that can lead to chronic degeneration of both the nervous system and the vascular system with serious effects on health and mortality.

Whereas Figure 1.8 focuses on the centrality of the pancreatic islet hormones, insulin and glucagon, as competing co-regulators of blood glucose, there are major control contributions from the central nervous system (CNS). There is, however, a growing appreciation for the pervasive and long-term role of the CNS by collective processes termed "interoception" in both the short-term (minute-to-minute) and long-term (year-to-year) metabolic regulations (Mirzadeh et al., 2022).

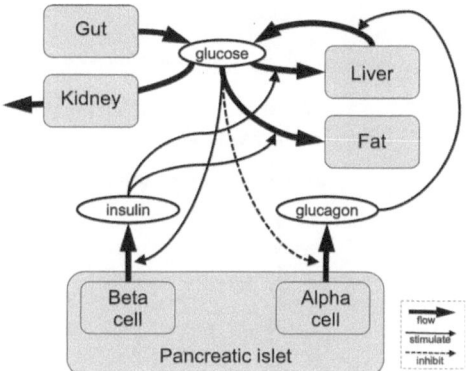

FIGURE 1.8 System map showing the amount of glucose in the blood results from (1) absorption of glucose by the gut, (2) excretion by the kidneys, (3) reversible storage as glycogen by the liver, and (4) as fat in adipose tissue. These processes are adaptively controlled by the hormones insulin and glucagon as are released in a reciprocal manner from beta cells and alpha cells of pancreatic islets.

Cellular Metabolic Energy Production and Regulation

From the above, we can see that DM is a disease defined and treated in terms of molecules of glucose and molecules of insulin. However, the disease is a multiscale, multidomain problem involving intricate pathways of intracellular metabolic processing, electrophysiology of membrane ion fluxes, the distribution of biochemicals among participants in multicellular metabolic pathways, and, ultimately the pathological degradation of the nervous system and the circulatory system.

Glucose concentration in the blood is tightly regulated to about 5 mM (or 100 mg per 100 mL of blood serum) in a feedback control loop by which increased blood glucose triggers pancreatic beta cells to secrete insulin into the blood by which insulin accelerates the uptake of glucose from the blood into most other cells in the body.

This pervasiveness is a consequence of the fact that cellular processes – e.g., cell replication, contraction, hormone secretion, and cell–cell signaling – all depend on metabolic energy production which, in most cases, is a process whereby metabolic fuels such as glucose (Figure 1.9) and fats convert low-energy ADP (adenosine *di*phosphate) to high-energy ATP (adenosine *tri*phosphate). This chemical energy fuels all manner of muscle contraction, gene and cell replication, hormone biosynthesis, and signaling. These basic metabolic pathways are central to understanding how glucose regulates insulin secretion from beta cells and for understanding how insulin controls the uptake, storage, and use of glucose as an energy source in fat and muscle cells.

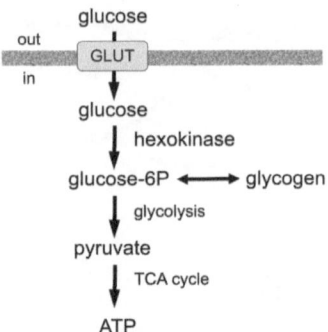

FIGURE 1.9 Cellular pathway by which extracellular glucose is transported into the cell by the membrane glucose transporter (GLUT) where it is phosphorylated to glucose-6-phosphate (glucose-6P) which enters the glycolysis and tricarboxylic acid pathways to produce ATP which is the cellular energy currency that fuels all manner of other cellular processes.

Cell cytoplasm (Figure 1.10) is a proteinaceous gel solution that contains solutes such as inorganic ions (Na^+, K^+, Ca^{++}, Cl^-, PO_4^-, etc.), dissolved gasses (O_2, CO_2), small molecules (carbohydrates, amino acids, peptides, etc.), and large proteins, some diffusible and others immobilized, that serve as binding site buffers for other constituents or as enzymes. The cell membrane that separates the cytoplasm from the surrounding extracellular "matrix" is a lipid bilayer that is, generally, impermeable to large or ionized atoms and molecules. The cell membrane is embedded with membrane proteins that can transport material and information into and out of the cytoplasm. For example, membrane receptors bind certain extracellular molecules, change their conformation on the inside of the cell, and thus affect an enzymatic process that it catalyzes. Other membrane-bound transporters are "ion-pumps" that can transport ions and molecules into or out of cells.

Cytoplasmic organelles perform specialized biochemical and genetic tasks such as: cell nucleus contains the cell's genes and DNA-to-mRNA transcription machinery, ribosomes that synthesize proteins according to messenger RNA (mRNA) sequences, endoplasmic reticulum that packages newly formed proteins for transport in the cell, mitochondria generate biochemical energy as ATP, and microtubules are long protein polymers that guide material transport throughout the cell.

This cast of cellular material entities (plus others not listed) participates in a wide variety of interlinked biophysical processes in networks that are often diagrammed as below. However, such diagrams and their textual

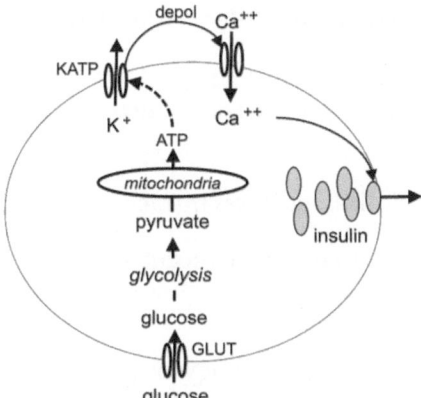

FIGURE 1.10 The secretion of insulin secretory granules (right) depends on the uptake of calcium ion (Ca++) through depolarization ("depol") dependent calcium ion channels. Membrane depolarization depends on the closure of ATP-sensitive potassium ion (K+) channels (KATP) that depends, in turn, on the metabolic production of ATP from glucose as taken up through glucose transporters (GLUT).

interpretations are extreme simplifications of all the entities and processes that are operative in a beta cell at any one time. It is precisely the challenge of building computational representations of biological process networks that have motivated our work.

Beta-cell insulin secretion can also be inhibited by the peptide hormone somatostatin, and by sympathetic nervous system inputs. Each cell type – e.g., alpha, beta, and delta cells of the pancreatic islet – plays a distinct physiological role in the network of organs by virtue of the specific molecular components that compose the cell and the molecular signaling pathways by which the cell operates.

For example, if one considers the causal pathways diagrammed in Figure 1.10, one can articulate qualitative causal pathways consisting of discrete process steps by which glucose controls insulin secretory rate:

- *Increasing* extracellular glucose concentration *increases* glucose transport into the cell.

- *Increasing* intracellular glucose metabolism *increases* the production of ATP.

- *Increasing* ATP/ADP ratio *reduces* membrane K_{ATP} potassium ion channel activity.

- *Reducing* K_{ATP} channel activity *reduces* (i.e., depolarizes) the cell membrane potential.

- *Reducing* membrane depolarization *activates* membrane calcium ion (Ca^{2+}) channels.

- *Increasing* Ca^{2+} influx *increases* intracellular Ca^{2+} concentration.

- *Increasing* intracellular Ca^{2+} concentration *increases* the exocytotic release of insulin.

Understanding the cellular physiology of pancreatic cells and how they regulate systemic metabolism is a topic of years of diabetes research into the complex set of interacting biochemical and biophysical processes within the islet. In order to explain normal and diabetic insulin secretion, investigators have had to consider – i.e., to rule in or out – an amazing number and combinations of physiological features. A partial list would include the pathways in Figure 1.10:

- Three islet cell types that synthesize and secrete a specific peptide hormone: beta cells secrete insulin when blood sugar level increases;

- Each islet hormone has a specific metabolic effect, often antagonized by the other islet hormones, on virtually all cells in the body including fat, muscle, and nerves. Most germane to diabetes and metabolism is that insulin increases the uptake of glucose into cells;

- Islet hormones have both long-loop feedback and short-loop paracrine effects to stimulate and inhibit hormone secretion from the other islet cell types.

Systemic Control of Metabolism

The CNS – brain, hypothalamus, brainstem, spinal cord, and spinal nerves – has ultimate authority over metabolic processes. This includes the control by hypothalamic nuclei of body weight by controlling hunger, food intake, and physical activity. It includes the (usually) competing effects of sympathetic and parasympathetic branches of the autonomic nervous system to regulate pancreatic hormone release, hepatic metabolism, and metabolism in fat, muscle neural tissues.

Disease as Failure of Feedback Control

A central challenge is that each disorder is a failure of physiological control systems to constrain blood pressure and blood glucose concentration to normal, non-pathophysiological levels. Consequently, these diseases are analogous to failures of a thermostat to control room temperature to a setpoint temperature or of a car speed controller to maintain a setpoint speed.

However, such analogies are flawed in a manner that has profound implications for understanding physiological control systems. Technology-based control systems are designed to compare a physical measure of the controlled variable (e.g., temperature or speed), compare that to a "target" value, and use the difference to generate a feedback effort that normalizes the controlled values. However, physiological "control" systems have no such explicit representation of their target values. For example, there is no sample of fluid that serves as a pressure standard for normal blood pressure, nor is there a quantity of chemical that serves as a concentration standard for a normal glucose level.

CHALLENGES OF MULTISCALE, MULTIDOMAIN ANALYSIS

In the foregoing, we have very briefly described two outstanding public health problems that result from failures of proper feedback control of critical physiological control systems. As such, both failures present challenges to representing, analyzing, and understanding dynamical systems. First, these clinical problems are "multiscale" in that they involve players at multiple anatomical scales ranging from atoms, molecules, cells, organs, organ systems, and onto whole organisms. Second, they are "multidomain" in that they involve all manner of physical entities and phenomena including physical processes in biochemical, fluid dynamical, mechanical, and electrophysiological domains.

Domain Technology "Silos"

Given the organismal complexity and homeostatic intricacies of both diseases, empirical and analytical investigators must necessarily focus on particular phenomena at particular structural levels and temporal spans. Inevitably, what is learned about a particular physiological phenomenon or pathological outcome becomes locked into intellectual and technological "silos" that often correspond to a specific academic and clinical domain – biochemistry, embryology, genetics, etc. Silos can center on particular technologies (electrophysiology, gene sequencing, X-ray crystallography, etc.) that necessarily develop their own data types and conceptualizations or can be based on the structural scales and physical domains. The result can be terminological and conceptual barriers enclosing scientific "towers of babel" or "silos" that impede common understandings, vocabularies, and data sharing for solving broader-scale problems.

Such knowledge silos are accentuated by the evolution and use of incompatible data standards, computational languages, and terminological conventions. These issues are of particular concern to the mathematical modeling community that has developed out of many scientific and technological disciplines so that the National Institutes of Health has funded the Center for Reproducible Biomedical Modeling (https://reproduciblebiomodels.org/) whose charge is to "enable larger and more accurate systems biology models, as well as their applications to science, bioengineering, and medicine, by enhancing their understandability, reusability, and reproducibility".

These are the problems being addressed by the biomedical informatics community as they have developed standard terminologies, data formats, and modeling languages. To provide a deeper, more formal level of meaning to mediate knowledge exchange, formal ontologies have emerged since the mid-1990s as exemplified, at the molecular level, by the Gene Ontology (Ashburner et al., 2000), at a gross anatomical level by the Foundational Model of Anatomy (Rosse & Mejino, 2003), and for physical systems by the PhysSys ontology (Borst et al., 1995). These were followed by the establishment of biomedical ontology repositories such as BioPortal (Noy et al., 2009) and Open Biomedical Ontologies (OBO, 2006). These resources are well described in the next chapter about bioinformatics resources.

The Systems Perspective

The overriding challenge presented by these common diseases, and by virtually all other organismal processes, is the sheer number and complexity of participating physical entities and the numbers and kinds of processes by which they participate and interact. Comprehensive analyses of large-scale and complex systems require a physics-based, "systems dynamics" approach as has emerged from the engineering sciences and adapted biophysical analysis. Such methods are particularly effective, even essential, for the quantitative analysis of invented structures and whose parts can be closely and quantitatively observed. However, the complexity of biophysical structures, functions, and processes can be expressed and analyzed in a similar quantitative manner. However, workers are challenged to gather sufficiently accurate quantitative data to formally test such hypotheses.

We thus present an approach to describing and modeling physiological systems that makes no distinction between what is normal versus what is abnormal. At the levels of chemicals, cells, organs, and organ systems, it is all a matter of the laws of physics.

Challenges of Feedback and Homeostasis

Biological organisms have an inherent ability to self-regulate their processes and states within narrow physiological operating ranges so as to assure health, survival, and reproductive capacity. Pancreatic beta cells rely on the kidneys to maintain serum sodium and chloride levels in a range that supports normal membrane electrical activity. They rely on the liver, pituitary glands, and kidneys to maintain proper blood osmolarity. Furthermore, the hormone levels surrounding beta cells must all be in a proper array to assure that insulin secretory rate is consistent with the hormonal, metabolic status. Such imperatives hold for all cells in the body in some frankly mysterious way.

> Claude Bernard (1813–1878) originated the term *milieu interieur* to describe the stability of physiological properties and processes at a time when physiology was still a science of grossly observable phenomena. Later, Walter Cannon (1871–1932), in his book The Wisdom of the Body (1932) generalized this idea as "homeostasis" to describe the processes by which systems maintain a milieu interieur of normal, stable property values such as normal blood pressure, body temperature, and blood nutrient level. Such observations raise key questions about how such stability emerges from such a collection of anatomical entities as they participate in a myriad of physiological processes.

Whereas Bernard identified a "milieu interieur" and Cannon postulated that homeostatic processes are at work, one is left with a vision of whole-organism physiology as a vastly interconnected network of ongoing processes and feedback effects wherein the rate of each process depends, to some extent, on the rates of each of the other processes. This is a rather daunting picture as a vastly complicated analytical problem but it reassures that we can focus on particular events –blood sugar or pressure levels, for example – assured that by whatever mechanism the existing *milieu interieur* assures that key physiological conditions provide a stable computational state for modeling. For example, a model of insulin secretion or of blood pressure regulation can assume a normal concentration of calcium ion in the extracellular space as "boundary value" that stands in for one aspect of the *milieu interieur* of the pancreatic islet.

Disease as Feedback Control Failures

A central challenge is that each disorder is a failure of physiological control systems to constrain blood pressure and blood glucose concentration to normal, non-pathophysiological levels. Consequently, these diseases are analogous to failures of a thermostat to control room temperature to a "setpoint" or "target" temperature or of a car speed controller to maintain a setpoint speed. Such systems compare a physical measure of signal (e.g., a room temperature or a vehicle speed) to a target value variable (e.g., a temperature or speed), compare that to a "target" value, and use the difference to generate a feedback effort that normalizes the controlled values. However, physiological "control" systems have no such explicit representation of their target values. For example, there is no sample of fluid that serves as a pressure standard for normal blood pressure, nor is there a quantity of chemical that serves as a concentration standard for a normal glucose level.

Failures of Feedback Control

The failure of regulatory systems to maintain the normal glucose level accelerates the incidence and severity of atherosclerosis, heart disease, brain strokes, eye damage, and nerve damage (diabetic neuropathy). We will focus on the physiological and pathophysiological aspects of type 2 diabetes as clinically important failures of normal performance of a feedback regulatory system that is multiscale. This feedback system involves process participants such as transmembrane fluxes of ions (e.g., Ca^{++}, K^+, and H^+), complex cellular metabolic, genomic, and secretory processes, as well as neuroendocrine signaling. The main problems are in how insulin-secreting beta cells of the pancreas participate in complex cell signaling networks involving cells in adipose (fat) tissue, muscles, and liver.

Diabetes involves changes in multiple entities and processes in complex causal networks that aim to account for functional disruption of insulin-secreting beta cells, their failure to properly secrete insulin, and the failure of metabolizing tissues to properly respond to insulin. The disruptions of the causal networks responsible for maintaining normal glucose levels in the blood ("euglycemia") involve multiple cell types in the islets of Langerhans of the pancreas, liver cells, fat cells, muscle cells, and metabolic regulatory cells in the hypothalamus of the brain.

As with modeling cardiovascular function, there is a long history of dynamic modeling of the endocrine-metabolic control of blood glucose in health and disease. Comprehensive reviews include the evolution of such models ranging from the so-called "minimal model" of Bergman (Bergman, 2021) to multidomain, multiscale models (Ajmera et al., 2013) that span from organ-level models to molecular–cellular models as illustrated below.

Finding Feedback Pathways

Whereas blood pressure control may behave as if it is a classic feedback control system, there are features of biological control systems that can defeat the analytical methods of automatic control engineering. Linear controls theory has been extended and elaborated into the digital age to control the flights of airplanes and rockets and to drive our cars. Yet even these problems do not rise to the level of hierarchical complexity of most biological control systems. Blood pressure is regulated by multiple parallel feedback loops operating on several different effectors that control heart rate, stroke volume, vascular flow resistance, fluid excretion by the kidneys, and fluid intake. Each of these control paths operates on effectors that themselves have nonlinear control properties including hysteresis. One of the more profound problems in using engineering controls theory is that in biological control systems there is no "setpoint"; rather, setpoints are emergent properties of the entire system that have been determined by gene selection.

A classical control system aims to regulate a controlled variable so that it matches a setpoint value. A furnace controller matches room temperature to the setpoint temperature as set by a dial. An automobile speed controller matches the speed of a car to the setpoint value as set by the driver. However, there are no such explicit physiological setpoints for blood pressure and none for blood glucose level. For example, there is a container of fluid pressurized to a setpoint pressure or a quantity of fluid with a precisely normal concentration of glucose that serves as a normal setpoint for blood glucose levels. Whereas one might posit that physiological setpoints are somehow, themselves, a controlled quantity, but this simply displaces the question to "What sets the setpoint?"

However, we do find it useful to invoke the term "control system" to describe the observed behaviors of physiological systems because they behave "as if" setpoints and feedback loops were in play. Indeed, just this approach is extremely valuable if one needs to implement a technological controller to replace or aid an otherwise missing or injured physiological system. Such artificial control systems are essential for "glucostats" that automatically control blood sugar in diabetics by supplanting the mal-functioning pancreatic islets with insulin injections to maintain a glucose setpoint.

Whatever the actuality of physiological setpoints and feedback control, hypothesizing the existence of a feedback controller is a useful tool for characterizing the functional structure and operating characteristics of an observed physiological system. Certainly, the so-called "normal" values invoke, at least implicitly, an operating feedback control system. For example, it is useful to describe a normal "setpoint" blood glucose concentration or for blood pressure. Beyond that, it is useful to hypothesize a simple analytical feedback model to fit dynamical data to establish normal values as operating setpoints, feedback gain, and response dynamics. We will describe two medically important feedback control systems – one that controls blood pressure, the failure of which is called "hypertension", while the other controls glucose concentration in the blood (hence, "blood sugar"), the failure of which is "diabetes".

Need for Informatics and Analysis

In the foregoing, very brief description of the physiology and pathology of diabetes and hypertension, we have emphasized the broad scope and depth of the anatomical and physiological systems that are role players in normal and disease physiology. Figure 1.2 shows important aspects of these systems by illustrating (1) the wide range of physical things that are participants (atoms to bodies) across a broad range ($\times 10^{13}$) of sizes, (2) the classification of these things according to types, and (3) the broad range of processes organized by time scale.

Classes of Things – Ontological Continuants

We have described clinical problems that present multiscale, multidomain challenges to studying, modeling, and understanding normal and pathological processes in health and disease. We start by noting the breadth of spatial scale ($\approx 10^{13}$) of biological things that range from atoms to bodies of organisms that are of concern to biomedical and physiological investigators (top panel). Furthermore, within each class of thing, there are many subclasses that range from 2 body types (e.g., male, female) to well over 100,000 different molecule classes (e.g., from ions to proteins and genes) that must be identified, characterized, and named.

These processes occur across the many physical domains and on time scales ranging from atomic motions and molecular processes to the disease progression and lifetimes of whole organisms. Furthermore, there is a comparably broad range of temporal durations of physiological processes that roughly parallel the range of spatial scales.

Classes of Processes – Ontological Occurrents
Even more daunting is the challenge of naming, defining, and counting multidomain, multiscale processes. For example, one could define a simple model of blood pressure regulation as some algebraic or differential equation relating the value of blood pressure to heart rate, and then elaborate the model by including heart rate and ventricular contractility, and then by a more detailed model of myocardial muscle mechanics. The point is that process definitions are a daunting, open-ended combinatorial problem because each participant in a process depends, in turn, on the participants in one or more other processes.

Physiological Domains
The anatomical and physiological scope of bodily systems and processes that participate in blood pressure and blood sugar regulation involves physiological processes across several biophysical domains. Such multidomain problems are challenging simply because they span traditional academic and disciplinary domains that are distinct knowledge "silos" that each have their own preferred vocabulary, jargons, data storage, analytical methods, and approaches to problem solving (Table 1.1).

The existence of such knowledge and technology domain boundaries present real impediments for collaborative solutions to multidomain problems such as diabetes and hypertension because of field-to-field differences in conventions for, among others: (1) anatomical naming, boundaries, and terminologies, (2) units and scaling for measurement data, (3) theoretical abstractions for parts and processes, and (4) methods for data

TABLE 1.1 Multiple Biophysical Domains Relevant to the Scientific Study and Treatment of Human Disease

Knowledge Domain	Physiological Processes
Fluid mechanics	Blood flow, air flow
Solid mechanics	Heart contraction, skeletal motion
Electrophysiology	Nerve signaling, muscle excitation
Chemical kinetics	Enzymatic processes, gene transcription
Diffusion kinetics	Respiratory gas exchange, nutrient distribution
Heat transfer	Body temperature regulation

and information representation, storage, and retrieval. Whereas many of these knowledge representation differences can be sufficiently resolved for informal communication, computational and database problems are being resolved using bioinformatical tools (reviewed in Chapter 2) and biomedical ontologies (reviewed in Chapter 3).

HYPOTHESIS EXPRESSION AND ANALYSIS

So far in this chapter, we have described two complex use-cases of, as-yet, unsolved clinical problems that have motivated the development of powerful computational methods for representing, archiving, displaying, and analyzing hypothetical and consensus knowledge about outstanding biomedical problems such as diabetes and hypertension.

The integrated function of an organ, a heart for example, depends on the functional connectivity of its parts such as its ventricles, valves, and vessels, yet each connection consists of cells, cell parts, and molecules. The same holds for hormone secretion, and an enzyme molecule function depends on its catalytic site and the binding sites for its modulating signals. There is also functional information flow between layers. The environmental demands on organs such as an increase in musculoskeletal workload signal changes in lung function and muscle blood flow that, in turn, create demands and adjustments among the cells constituting these organs and organ parts. The molecules and molecular systems then respond, and these changes are propagated into the cell layer and on to the organ layer.

Narratives, Diagrams, and Computations

In the foregoing, we have used verbal narratives and schematic diagrams to illustrate and reason about the functions of complex physiological systems. This is not a new idea, but it is one that only recently has garnered sufficient computational and informatics support to offer widespread research benefits. Without computational support, methods may rely heavily on intuitive appreciation of biophysical mechanisms such as appreciating that the flow of a chemical species through a synthesis pathway will drain the source pool of substrate and augment the product pool. Such reasoning may account for the rate-limiting kinetics of an enzymatic reaction step and for the subsequent disappearance of the product species.

Qualitative Functional Reasoning

We have used schematic diagrams and verbal narratives to describe and reason about complex biophysical systems. Indeed, the vast amount of model expression is based on informal or formalized icon-arrow diagrams using qualitative reasoning and assertions such as "if blood pressure increases, then blood flow rate increases".

As the scope and depth of our knowledge grows about biology, our ability to understand, analyze, and make use of this knowledge becomes severely stained. We are faced with representing and understanding a bewildering array of biological entities – hearts, cells, genes, metabolic products – and their functional interactions as they participate in biological processes at all structural scales from atoms to organisms and across temporal scales from microseconds to lifetimes.

Our goal has been to leverage available knowledge resources – ontologies, terminologies, databases – that represent biological things from whole organisms to molecules and their atomic parts, and then create a physics-based computational theory for representing the processes in terms of the measurable physical properties and attributes of the entities and processes. As comprehensive as these computational knowledge resources are, we see outstanding challenges for (1) data and knowledge reuse, reproducibility, and reasoning, (2) data annotation, archiving, and access, and (3) database access formats, modeling languages, platforms, and code access. In this chapter, we will present and discuss computer graphical approaches for representing biophysical structures and behaviors and then describe the various bioinformatical resources and approaches for representing, archiving, and accessing biophysical knowledge.

As discussed above, icon-arrow system maps are readable visual, qualitative guides to the structure and function of complex biological processes. Even if approximate and sketchy, they are very useful for initial brainstorming and discussions of hypotheses. When formalized in terms of well-defined icons and arrows, they can be the basis for qualitative reasoning and hypothesis inference about behaviors and outcomes.

Quantitative Analysis and Simulation

Diagrams and verbal statements such as those presented above are the first phase of expressing, discussing, and analyzing biophysical hypotheses. However, the gold standard for functional reasonings relies on the design, implementation, and analysis of physics-based quantitative models. Such models are based and coded using the principles of classical physics as adapted and specialized for biological entities and systems.

A mathematical simulation model of such a system is written entirely in terms of variables that represent the *quantitative* values of physical properties. For example, the mathematical model code may declare variables

"X1" and "X2" to be in units of moles, and "R12" to be in units of moles/ second. However, the biological meanings of the variable can be known by some form of explicit annotation that X1 and X2 refer to the chemical con- centrations of cytoplasmic *glucose* and *glucose-6-phosphate*, respectively, and that R12 is the reaction rate cytoplasmic enzyme *glucokinase* accord- ing to some reaction rate equation. As will be more thoroughly developed in Chapter 5, which will describe early Nobel-prize-winning derivations in the 1950s by Michaelis and Menten of their model simulating enzyme mechanisms and by Hodgkin and Huxely of their model ion channels acti- vation and deactivation.

CHALLENGES – NEED FOR INFORMATION, KNOWLEDGE, AND ANALYSIS

The problem we face is that even the most basic physiological systems and their devastating diseases – diabetes and hypertension for our examples – involve entities and processes that span structural scales, physical phe- nomena, and "omic" domains. The challenge is to develop methods for representing hypothetical models and the analytical tools to manipulate and test hypotheses.

As we will describe in the following two chapters, the past few decades have witnessed the development of several computational technologies designed for representing the vast amount of data and knowledge needed for mod- ern science and technology. For example, the World Wide Web (WWW) depends on the idea of Universal Resource Locators (URLs) that serve as unique identifiers for representing particular computational or information entities. Translated into practice for biomedical and bioscientific uses, such *machine-readable, unique identifiers* were applied to the generation of data- bases, terminologies, and ontologies to enlist, categorize, and interrelate all of the bits and pieces of biomedical and biophysical knowledge.

As we catalog in the next chapter, the ability to provide unique, machine- processable identifiers has led to an explosion of resources (including ontol- ogies) describing biological physical entities (e.g., anatomy) and biological processes. Our focus is on developing computational, physics-based mod- els of complex biophysical systems. The aim of such models is to (1) iden- tify and represent each of the physical entities (e.g., molecules, cells, and organs) that participate in system behaviors that are of interest; (2) identify and represent the physical processes that account for observed behaviors; and (3) propose and implement computational representations of the pro- cesses and participants in terms of the mathematical and physical laws.

CHAPTER 2

Biomedical Information and Data Resources

I F OUR GOAL IS to model physiology, then what "stuff" about biology and physiology do we need to know? In this chapter, we describe the range of knowledge and information needed to effectively model physiology or pathophysiology. Fortunately, as the field of biology has become more information and data-driven, scientists have developed a host of relevant and public bioinformatics resources. Here, we provide an organized description of resources that are relevant from our physiological perspective. We conclude this chapter by providing some synthesis and presenting some prior research developing "Physiome"-style projects.

As we list and describe these resources and projects, questions about how the information is organized and stored will arise. These questions are part of larger, more theoretical, and philosophical questions about entities and processes. Although this text aims to be pragmatic, we will need to describe some fundamental aspects of *ontology* and *knowledge representation*, as these are central to the design of the Ontology of Physics for Biology (OPB) and to our physiological modeling approach. However, for now, we postpone those discussions for Chapter 3, as well as a definition of the word "ontology". We feel that readers should first understand the stuff, before thinking about philosophies about how to describe and organize that stuff. Nonetheless, even an informal survey of knowledge resources necessarily includes the term "ontology", and we will quickly encounter different ways to organize the stuff of biology.

DOI: 10.1201/9780429469961-2

One example of an important theoretical distinction is the difference between representing and storing information about *physical entities* versus *processes*. We will more formally define these concepts later in Chapter 6, but for now, we can rely on the intuitive notion that *physical entities* are things that occupy some region in space and can participate in *physical processes*. However, physical entities are describable without reference to what they do (their functions) or the processes in which they may participate. In contrast, *processes* occur over time – they have a start time and a stop time and a set of physical entities that are their *participants*. A process *is* a *change*. It is a change such as a flow of energy or matter. One example might be a change in the volume of some compartment (such as a squeezing heart ventricle). The attributes by which one notices or measures these changes are *physical properties*, such as mass, volume, electrical charge, or chemical concentration.

With these ideas, we can organize the sort of entities available in current bioinformatics resources. We will start our review with physical entities, going from large (biological organs) to small (proteins and small chemicals). Next, we consider resources about biological processes, and finally, we look at resources that include knowledge about properties.

BIOINFORMATICS RESOURCES: PHYSICAL ENTITIES

The largest sort of physical entity we might consider in the context of physiological modeling is the species or organism under consideration. For many applications, we are concerned about physiology and pathology in *Homo sapiens*. However, biology has a long tradition of working with "animal models", ranging from single-cell species (such as *E. coli*) to mammals such as mice or monkeys. Thus, many data resources consist of results from experiments or observations on such animal models.

One high-level resource is the NCBI Taxonomy (https://www.ncbi. nlm.nih.gov/taxonomy). As this resource is part of the National Center for Biotechnology Information, this listing and organization of species is described as covering species that are included in "the public sequence databases". Although this is a small fraction of all life on the planet, it does include over 400,000 species, and certainly all of those of interest to biomedical researchers.

Research into pathophysiology such as diabetes or hypertension (as in Chapter 1) often involves experimentation on these animal models. These experiments may concern large-scale physiology, such as the heart rate of mice that are genetically predisposed toward hypertension, or much smaller-scale physiology, such as understanding subcellular metabolism and glucose, which is relevant to diabetes. In many cases, although the experiments may be carried out on mice or in yeast cells, the research often claims that results generalize to other species. For example, it is reasonable for researchers to propose that physiological findings from mice apply to all mammals, once scaled appropriately for differences in overall mass. As a result, taxonomic information is sometimes vague (e.g., for basic cellular metabolism) and the physiological processes are often the same for all eukaryotes, rather than specific to a particular species.

Anatomy Resources

For a given species, the largest physical entities are the structural components of the anatomy of that organism. For the hypertension use-case of the previous chapter, the anatomic entities of interest might be things like parts of the heart, specific veins and arteries of the circulatory system, and the fluid contained in those vessels. For a study of diabetes, the entities might be the pancreas, the liver, and cell types such as beta cells or epithelial cells.

Anatomy is an ancient field of study, but only recently have these efforts been captured in a systematic and computable manner. One of the most important systematic efforts in anatomic representation is the Foundational Model of Anatomy (FMA; Rosse & Mejino, 2003). The FMA is a comprehensive ontology and resource for the organization of structural anatomical concepts for the human body. The FMA provides a static view of anatomy; it intentionally excludes all functional definitions of entities and any descriptions of pathology or physiology. Thus, the development of the FMA led directly to our thinking about how to similarly develop a systematic and principled view of physiology. Because the FMA is so central to our development of the OPB, we will return to a more detailed description of the FMA resource in the next chapter.

Of course, the FMA is not the only bioinformatics resource for anatomy. For example, a species-specific view of anatomy is provided by the Mouse Adult Gross Anatomy ontology (Hayamizu et al., 2005). An anatomy ontology that is specific to a particular model organism is important when considering anatomic features that may not be the same as for *Homo sapiens* (for example, the location of the mammary glands). There are quite a few resources for species-specific anatomy. For example, the zebrafish is a valuable and well-studied animal model for many reasons (including transparency); as with mouse anatomy, there is a zebrafish anatomy ontology (van Slyke et al., 2014).

For both of these model organisms, there are resources that catalog both *adult* anatomy and *developmental* anatomy. In general, if one wishes to understand pathophysiology, one often needs to understand both pathology (in the adult) and developmental processes that may have led to (or affected) that pathology. Thus, separate from the adult mouse anatomy ontology is the Mouse Anatomy and Development Ontology (Hayamizu et al., 2013), which describes anatomy as it develops through 26 mouse development stages. Similarly, the zebrafish resource includes information about both adult and developmental anatomy.

Although a plethora of resources provides for an excellent coverage of biological anatomical concepts, it leads to problems, especially if one wants to integrate data across species. Different resources organize and name anatomic entities differently. Sometimes, these differences relate to real differences across species, but sometimes the differences are more arbitrary. For researchers who need to integrate data and knowledge across these resources, they must either (1) develop mappings that connect the resources or (2) develop and use a single overarching anatomy ontology that links in a formal, principled manner the different resources. The latter approach is espoused by the developers of *Uberon*, an integrative, multi-species ontology (Mungall et al., 2012).

Uberon aims to generalize across species, so that other classes from other ontologies can be logically connected. It also aims to include developmental biology, such as mouse developmental stages. It does not aim to replace or be a superset of all anatomy ontology, but instead includes classes only when there could be a plausible generalization across species. Thus, Uberon does not include the depth or details of any single-species anatomy ontology.

As we will see throughout this chapter, a recurring problem we face is that of integrating data and knowledge across different resources. Although Uberon is one approach to addressing this problem with respect to anatomy, it does require significant development effort, including maintenance as individual anatomy resources may change over time.

Historically, anatomy describes entities that are structural and visible. But should this include microscopic anatomy? What about the anatomy of a cell, or the structure of DNA? As we show below, it can be a challenge to understand the boundaries of "anatomy" and to know which resource to use for information about microscopic entities.

Tissue and Cell Type Resources

Cell types and cell structures are important sources of information for both normal and pathological biology. An intuitive example is that the functions and processes in a neuron are very different from those in a pancreatic beta cell. Of course, all cells share some anatomy and some physiology, just as all mammals share some anatomy and physiology. When researchers need to document the cell type, many use the Cell Ontology (Diehl et al., 2016; Meehan et al., 2011).

As we describe in the next chapter, the Cell Ontology is part of the *Open Biological Ontology* (OBO) foundry collection of resources. The OBO aims to encourage ontology developers to follow a common set of principles with the goal of improving collaboration across ontologies. Thus, the Cell Ontology includes built-in connections to gross anatomy (via Uberon) and the Gene Ontology (GO) specification of biological processes (see below).

The Cell Ontology is designed to be pan-species, including cell types from prokaryotes to mammals, but excluding plant cell types. Thus, it includes terms such as "flatworm hypodermal cell", as these are distinct from vertebrate cell types. The Cell Ontology also includes cell types that are quite generic, and ones that simply refer to function, rather than structure. Thus, the ontology includes classes for "barrier cells" or "contractile cells", without specifying additional information about the cell structure or anatomic characteristics. As we describe in later chapters, we argue for a more careful, principled distinction between ontologies that describe the structure of physical entities (such as the FMA) versus the processes and functions that describe how those structures might behave over time.

Groups of cells of particular types can form coherent biological *tissues*. This class of biological stuff is important for experimentation that aims to understand genetic expression and characteristics when the sample size is of the granularity of these groups of cells. For example, an experiment may test the behavior and/or genetic expression of cardiac muscle tissue (the myocardium). The Brenda Tissue Ontology (the BTO, Gremse et al., 2011) also includes fluids, such as blood, amniotic fluid, and cerebrospinal fluid. Of course, the constituent parts of these fluids and tissues are individual cells of particular types. Unfortunately, the BTO does not include direct links back to the Cell Ontology, as one might hope. As we describe later, many ontologies are not as connected as they should be, which leads to challenges when using more than one ontology.

Cellular Components

Smaller than individual cells, consider the parts of a single cell – cellular anatomy. For these concepts, the most important resource is the "cellular component" portion of the GO. The GO is one of the oldest and most successful biological ontologies and is divided into three sub-ontologies: biological process, molecular function, and cellular component (Ashburner et al., 2000; Carbon et al., 2019). The cellular component tree includes terms that represent structural subcellular anatomy, such as organelles (e.g., the ribosome or the mitochondrion), membranes, and regions such as the nucleus and cytosol. It also includes macromolecular complexes that play key roles in subcellular processes. The GO class, "protein-containing complex", is defined very broadly. However, in practice, the GO children of this class are complexes that carry out specific activities that are found within the GO-molecular activity tree; it does not purport to include all stable biochemical complexes found within the cell. An alternative resource specifically for macromolecular complexes is the Complex Portal (Meldal et al., 2019). This resource is manually curated and also does not claim to be comprehensive, but as of 2023, included information about more than 3,800 complexes.

One could argue that subcellular components such as the organelles and membranes are clearly structural and, thus, might be more properly described as the smallest parts within a comprehensive anatomy ontology such as the FMA. (Indeed, the FMA does include subcellular organelles.) However, where to draw the line between anatomy and chemistry is challenging: is a macromolecular complex part of cellular anatomy or biochemistry? After all, even specific chemical ions have a physical structure at the atomic level. However, because our goal is to understand the subcellular biochemical processes of many pathologies (e.g., cancer), it is important to catalog and include knowledge of physical entities and processes that include molecular-level biochemical interactions.

Proteins

An important subclass of biochemical entities are proteins – macromolecules composed of amino acids. These physical entities play critical roles in cellular behavior, cellular processes, and cellular pathologies, which ultimately can lead to disease at the organism level. Because of this chain from microscopic to whole-organism pathology, if our task is to model physiology, in some arenas, these models include biochemical reactions among sets of proteins and other small biological molecules.

As is well known, data and knowledge about genomic science have exploded in recent decades, and the catalogs of genes, proteins, and metabolites are vast and rapidly growing. We cannot attempt to provide a comprehensive catalog of these resources (see, for example, the annual NAR database issues). Instead, in this chapter, we will highlight only a few of these, with the aim of elucidating clear definitions of the sorts of players (physical entities) and biological processes (e.g., biochemical reactions) that are important for modeling physiology.

A primary, well-used resource for protein information is UniProt (Bateman et al., 2021), a comprehensive catalog of proteins across many species. Uniprot includes sequence information for proteins, which provides enormous detail about the particular protein. However, this can also be a challenge – proteins are often named and characterized by the roles they play in particular subcellular processes. For example, a protein would be named the same across two different species if it carried out the same function, even if the sequence information differed across the species (and, therefore, the detail of the physical shape of that protein). Even within the same species, there are "isomers" that have different sequences, but carry out the same function in the same manner, and are therefore named as the same protein.

To avoid this problem, the Protein Ontology (PRO, Natale et al., 2017) is an alternative resource that provides an organization and hierarchy for *classes* of proteins. This resource groups proteins first into families of proteins that may be functionally related, then proteins that derive from a single gene, and then different forms of that protein that are sequence-specific. PRO also provides for organism-specific versions of a protein— e.g., human "interferon regulatory factor 5" (IRF5) versus mouse IRF5. As you would expect, PRO includes UniProt cross-references once the protein falls to the level of specificity of a single sequence.

In general, teasing apart physical shape and structure from behavior and function becomes difficult. This is most clearly apparent for the class of proteins known as *enzymes*. These proteins are defined as carrying out a specific function: catalyzing certain biochemical reactions. Especially for metabolic processes, enzymes are very well-studied, and there are several important information resources for enzymes. Importantly, enzymes are numbered and classified by the Enzyme Commission (EC) numbering system (Bairoch, 2000). Although all enzymes have an EC number, this number does NOT refer to the structure or physical organization of the protein that acts as a catalyst. Instead, the number refers specifically to the types of biochemical reactions that are catalyzed by the enzyme. In part because enzymes are neither consumed nor produced by these reactions, biologists often do not care about the physical structure of the enzyme, but rather about its functional characteristics (e.g., the degree to which the enzyme increases the reaction rate). Thus, a discussion of EC numbers belongs in our list of resources about biophysical processes (specifically, a biochemical reaction), rather than entities.

Small Chemical Species

The smallest physical entities that directly affect models of physiology are small chemical species such as calcium or potassium ions. These can be important players in some biochemical reactions (see next section), such as metabolic or signaling reactions that can affect the behavior and physiology of the cell, which can then affect larger biological tissues or the entire organism.

A comprehensive and well-organized resource for these entities is the Chemical Entities of Biological Interest (ChEBI) resource (Degtyarenko et al., 2008; Hastings et al., 2016). ChEBI includes (as of 2023) over 60,000 curated chemical compounds. Importantly, ChEBI not only includes chemical descriptions, but also organizes its entries within a relatively rich set of descriptive and functional classes. These classes are used by annotating chemicals with "has-role" attributes, so that one can say that a particular chemical "has-role" apoptosis inhibitor, or "has-role" analgesic, or "has-role" antibacterial agent, to give three examples. As with enzymes, these has-role annotations provide information about what the entity does, rather than what it is structurally made of.

For structural information, ChEBI leverages standards for molecular structural information; it includes both SMILES and InChI standards for specifying the structure of a single chemical compound. (SMILES or "Simplified molecular-input line entry system" is older and specifies a bit less information than InChI, the International Chemical Identifier standard.) Of course, this structured information is not available for more abstract classes of chemicals, such as the ChEBI term "polypeptide", which has many different polypeptides as children classes.

We have so far described resources that represent anatomical things across all structural scales but now we turn our attention to the representation of the sorts of processes in which such things participate in health and disease.

BIOINFORMATICS RESOURCES: PHYSICAL PROCESSES

In contrast to physical entities such as those described in the prior section, there are not as many information resources about physical *processes*. Yet, if we are studying physiology or modeling pathology, surely it is the processes that should be central. Drawing from the example in Chapter 1, what are the steps and processes that lead to the production of insulin? How can this temporal process be divided into smaller processes or named steps? As we discuss more in the next chapter, the intuitive notion of "part of" can be divided into two distinct sorts of relationships. First, a physical entity, like a mitochondrion, can be part of a larger entity such as the cell, and second, a simple physiological process like a biochemical reaction can be part of a larger process such as the production of insulin. Unlike physical entities, all processes have an attribute of *duration*. In this section, we will begin with the smallest (shortest) biological processes and build our way up to longer, more complex physiology.

Enzymatic Biochemical Reactions and EC Numbers

For physiological models, one of the most common building blocks for a model is a biochemical reaction. As with proteins, there are a wide variety of resources and standards for biochemistry, and we will not comprehensively cover these. The most well-studied and understood class of biochemical reactions are metabolic ones – those that are catalyzed via an enzyme.

Enzymes have been classified and categorized by the "Nomenclature Committee of the International Union of Biochemistry and Molecular Biology", and these were published (starting in 1992) as "EC numbers" (Bairoch, 2000). These EC numbers provide a classification according to the reactions that they catalyze. Thus, although most discuss EC numbers as if they describe an enzyme, which is a physical entity (a type of protein), we feel as though EC numbers are more accurately described as a classification of one sort of reaction, namely, those that have an enzyme as a catalyst. Importantly, if a single enzyme is "multi-functional", meaning that it can catalyze two or more different reactions, then that enzyme is given a different EC number for each of these reactions.

The ENZYME database (see https://enzyme.expasy.org/) provides a complete listing of EC numbers (~6,500 of them) and clearly specifies the reaction for each enzyme (Bairoch, 2000). This resource also includes cross-references to other reaction databases. Among these, the Rhea knowledge base provides more information about these reactions, including structural information about all participants (Alcántara et al., 2012; Bansal et al., 2022). Rhea includes almost all of the reactions with EC numbers and, in addition, includes *transport* reactions, which do not have catalyzing enzymes. Although Rhea includes descriptions of over 14,500 reactions, there is no claim that this is a complete set of significant biological biochemical reactions.

Protein–Protein Interaction Databases

A well-studied category of biochemical reactions are protein–protein interactions. In these experimental studies, two proteins of interest are studied *in vitro* to see if they might interact in any way, e.g., by binding to each other. This experiment is carried out in a high-throughput manner, so that very large databases can be assembled that represent protein–protein interaction networks. In these networks, a linkage between two proteins indicates that they do interact, and no link indicates that they do not. There are many protein–protein interaction databases that have different strengths and weaknesses (Lehne & Schlitt, 2009). Fortunately, there are several resources that unify and integrate information from other resources: e.g., GPS-Prot (Fahey et al., 2011) and Wiki-Pi (Orii & Ganapathiraju, 2012).

Although useful for initial qualitative assessments and screening information, these resources are not as useful for the sort of quantitative biosimulation models that is our focus. These networks provide no information about the participants of these reactions and no quantitative information about the interaction. Further, the *in vitro* information is only suggestive of what might happen *in vivo* in a particular cell type in a particular tissue. Because our aim is to model and understand the physiological mechanisms in sufficient detail so as to produce trajectories of events over time, we need significantly more information about proteins' participation in particular reactions in particular tissues.

Gene Ontology: Biological Process and Molecular Function

The GO has a long history as one of the most well-known and well-used ontologies in biology (Ashburner et al., 2000; Carbon et al., 2019). The GO consists of three distinct "branches", or hierarchies, of terminologies: Cellular Component (discussed earlier), Biological Process, and Molecular Function.

At a high level, the "molecular function" branch of the GO can be viewed as a type of biological process; but a very low-level process that can be characterized as the primary activity of a single gene product. The molecular functions of many enzymes are listed in this branch. The GO authors distinguished these molecular functions from broader, larger-scale processes such as "glycolytic process", which are included in the "Biological Process" hierarchy.

The primary goal of the GO is not to attempt a complete catalog of biological processes, but rather to provide consistent terms for annotating genes. The basic idea is that when one scientist claims a particular gene produces some protein that participates in a particular process (such as glycolysis), then they should use consistent terminology for that process. This allows for identification of homologs – genes across different species that carry out the same biological function, or are part of the same biological process.

The scope of "biological process" within GO was originally limited to cellular processes, or at most, communication among cells. Even after 20 years of expansion, the GO is primarily focused on subcellular processes, and mostly biochemical processes that could be characterized as a reaction or a set of reactions. As the name indicates, the focus of the GO is on genes – all terms within GO should be terms that could be used to annotate specific genes. Thus, although the GO currently includes a few large-scale biological processes such as "digestion" and "coagulation", terms are included only if there are annotations about specific genes and gene products that participate in those processes.

As we will describe in greater detail in the next chapter, the distinction between GO's "biological process" and "molecular function" is at least partially historical. Today, it is recognized that the word "function" is a bit of a misnomer, and instead, both these branches should be viewed as containing information about biological processes, albeit at different scales. For example, the GO term "Toll Binding" (GO:0005121) is a child of molecular function. It is defined for a gene X as being the function (or "activity") of binding that gene product, X to a toll protein (a transmembrane receptor). In a physics sense, we see very little difference between the biochemical process of binding X to a toll protein versus a larger-scale process such as transmembrane transport (where one sub-step of this process might be binding to a toll protein). Yet "transmembrane transport" is GO:0055085, a subclass of GO:biological process.

It is a challenge to uniquely name or even describe biological processes, which is why there are fewer of these resources. As we mentioned at the beginning of this chapter, an organizing principle for our physics-based approach is that physical entities are different from the processes that those entities participate in. The GO aims to provide terms that describe the processes a single gene might participate in. However, aside from the names ("Toll Binding") and an English definition, there is no formal definition of that process. As an example, one notable omission is any list of participants in that process.

Because the GO represents theories and knowledge about molecular-level biology, its information is necessarily incomplete. For example, even if gene X is annotated with the GO term "Toll binding", that does not necessarily imply that gene product X directly participates in a biochemical binding reaction with the Toll protein. Instead, for example, it may be that the gene product X up-regulates some other gene product Y, which in turn binds to the toll protein. The best that the GO annotation implies is that the molecular function "Toll Binding" is enabled in some way by gene X.

There are recent efforts to provide a more formal structure to the information in GO, including logical axioms that connect, where possible, lower-level "molecular functions" to larger-scale biological processes. With these additions, one can use the GO to retrieve richer information, potentially linking a set of molecular activities into a causal pathway, when one process might "positively regulate" (or negatively regulate) the next process (Thomas et al., 2019). Although this is a good step toward defining larger sequences of biological processes, the approach is limited by the underlying annotations and terminology of the GO. As mentioned above, GO annotations may not mean those products directly participate in a particular process. Additionally, even the more detailed processes in the "molecular function" subtree are still not at the level of including the biochemical reactions with lists of participants and quantitative information as we would need for a full biosimulation model.

As a distinctly different approach, other resources aim to enumerate and specify biological pathways from the ground up (we describe these below). Especially originally, the developers of the GO clearly separate their work from these pathway resources: a GO "biological process" is not a pathway, as it lacks information about participants or information about what subprocesses it might have as steps. As we describe next, although a "biological pathway" does not have a crisp definition, it does include information about how smaller-scale processes (e.g., biochemical reactions) are linked together as steps in some larger process that can be named as a pathway.

Pathway Resources

Subcellular biological pathways are often divided into two large categories: metabolic pathways and signaling pathways. The former are processes that have to do with either the use of energy to create larger macromolecules or the release of energy by breaking larger molecules. Effectively, metabolic pathways have to do with the energy needed to keep the cell alive. In contrast, signaling pathways describe processes that have to do with a cell's response to the environment. Thus, signaling pathways include the behavior of a neuron in response to the extracellular presence of GABA (gamma aminobutyric acid, an inhibitory neurotransmitter), or the response of an epidermal cell to the presence of the EGF (epidermal growth factor) protein. Signaling pathways, thus, cover a wider range of biological phenomena than metabolic pathways. At a representational level, these types of pathways can be treated the same – either type of pathway can be represented as a list or a graph of linked biochemical reactions, where the product (the output) of one reaction is used as the reactant (the input) to one or more subsequent reactions.

These pathways can rapidly become large and complex, and cataloging all pathways is a large task. An early effort at this was part of the Kyoto Encyclopedia of Genes and Genomes (KEGG) (Kanehisa et al., 2023; Kanehisa & Goto, 2000; Demir et al., 2010). These KEGG pathways were hand-curated and, in fact, hand drawn as graphs of reactions. Figure 2.1 shows the KEGG rendition of part of the mammalian target of rapamycin (mTOR) signaling pathway. Unfortunately, these diagrams do not conform to any standards, although the physical entities involved do include cross-references to some of the physical entity resources discussed in the

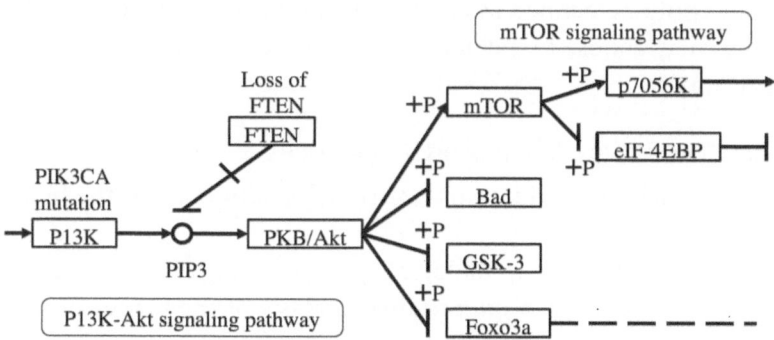

FIGURE 2.1 KEGG visualization of a portion of the mTOR signaling pathway.

previous section (e.g., UniProt and ChEBI). In general, KEGG resources are hard to align with other pathway and biological entity resources.

In recognition of the need for mapping pathway information across resources, the BioPAX standard was developed by a large consortium of users and developers (Demir et al., 2010). This format has been widely adopted and successfully captures the essence of information in any pathway resource. There are quite a few large, well-supported libraries and resources that list pathways and describe them using the BioPAX format. For example, BioCyc provides species-specific libraries of pathways that focus on metabolic pathways (Karp et al., 2019). In contrast, the Reactome resource includes signaling pathways and provides a greater level of detail, especially when describing protein-containing complexes that participate in signaling pathways (Gillespie et al., 2022; Vastrik et al., 2007). A significant success of the BioPAX standard is that information across many different resources can be syntactically compared.

Given that the goal of our research is to understand biological processes, it would be useful to be able to connect or link-up these low-level pathways into larger-scale processes. (We will say more about this idea; see the "Putting It Together: The Physiome Vision" section.) Unfortunately, even with syntactic alignment, it remains challenging to really connect or link pathways across resources. For example, even for well-studied and well-understood metabolic pathways, there can be significant differences in how these pathways are specified (Wang et al., 2016). These differences arise from differences in granularity, as well as disagreements about the boundaries of the pathway. Granularity differences can make the same reaction appear different, as the reaction may have more or fewer participants, or could be defined as several steps versus one. Likewise, different resources may choose to include or exclude reactions at the "boundaries" of a pathway.

Quantitative Biosimulation Model Resources

Pathway modeling provides *qualitative* information about the biochemical reactions and participants (proteins, small molecules, and cellular components) that make up biological processes. However, for most complex reaction networks, a *quantitative* analysis of the processes is essential for understanding behavior. The qualitative pathway information is very incomplete for any networks that include loops, and many critical biological processes involve positive or negative feedback loops. Without quantitative information about how fast a reaction occurs, one cannot predict when the effect of a feedback loop becomes dominating. This information is critical for researchers aiming to understand, for example, how a drug may disrupt (or not) a particular pathway.

In response, biological modelers have developed mathematical biosimulation models that provide fine-grained mechanistic explanations for biological processes and behaviors. The challenge, of course, is carrying out the biological experiments that provide the data needed to estimate these reaction rates. Thus, although there are thousands of published biosimulation models, these models cover a much more narrow scope of biology than the collection of pathway information. (One can know mechanistic details about a few things, or a broad sketch of the pathway about many more things.)

As with pathway models, there are standard formal languages for creating and defining biosimulation models. One of these, the Systems Biology Markup Language (SBML; Hucka et al., 2003), is the basis for a large collection of models available at the BioModels repository (https://www.ebi. ac.uk/biomodels/). This collection includes over 1,000 curated models, mostly subcellular models made up of biochemical reactions (both signaling and metabolic processes). In contrast, a different resource, the BiGG collection focuses exclusively on metabolic processes (King et al., 2016; see http://bigg.ucsd.edu/).

A broader range of models can be found at the Physiome Model Repository (PMR, Yu et al., 2011), where the modeling language is CellML (Lloyd et al., 2004). Like SBML, CellML is an XML-based markup language, but its focus is on the mathematics of the model, expressed using the MathML markup language. Nonetheless, these two languages are similar enough that the modeling community has begun efforts to make these two languages interoperable, so that information can be combined and compared across both repositories of models (Neal et al., 2019b). The PMR includes more than 600 models, with models ranging from skeletal mechanics, to blood circulation, to electrophysiology, as well as metabolic and biochemical reaction models.

Of course, researchers do not always use any standard languages for representing or implementing biosimulation models; models can be encoded in any programming language. Unfortunately, if researchers use general-purpose programming languages (such as C or MatLab), the result is usually that such models are not easily understandable, and their behaviors may not be reproducible.

In contrast to the pathway resources, all of these biosimulation models include sufficient mathematical detail that they can be executed or run by a simulation engine. The engine essentially applies time-step information, solves (or estimates a solution to) the mathematical equations, and produces as output, time-dependent traces of the variables of interest. This is a key capability: The ability to perturb an *in silico* biological system and then observe predicted effects on variables of interest could allow researchers to understand, for example, the potential effect of genetic mutations on specific processes, or the possible effect of a therapeutic agent on a pathological process.

PHYSICAL PROPERTIES FOR PHYSIOLOGY

But what are these variables? Ultimately, physiological and time-dependent processes are observed and *measured* as changes to particular physical properties of the participating entities. As a simple example, for tumor models, the growth in mass of a tumor is one property that can be measured over time. As a much smaller-scale example, the result of some pathway or biochemical reaction might be an increase (or perhaps an oscillation) over time of the chemical concentration of some protein or macromolecule in some region (e.g., the cytoplasm of a cell). Alternatively, the increase could be measured by an increased count (in moles) of that macromolecule in that region (thus, the variable might be measuring the same increase, but with different units).

Thus far, we have presented resources that describe physical entities, and then the processes in which they might participate. What about resources that name and list the types of physical properties that can be measured? In our view, this is one of the critical gaps that the OPB aims to fill. As we will describe in later chapters, a large portion of the OPB is dedicated to organizing and enumerating these physical properties. Here, we provide some background on existing attempts and related resources.

First, we acknowledge that physical properties are related to the notion of *phenotype*. After all, a classic example of phenotype is hair or skin color; and color is a physical property (it is measurable) of all visible physical entities. Sometimes, "phenotype" implies abnormal biological traits and properties. This is the focus of the Human Phenotype Ontology (HPO), which lists abnormalities across anatomic systems or cells (Köhler et al., 2021; Robinson et al., 2008). A challenge for this focus is in defining what "abnormal" stands in reference to. An intuitive example might be abnormal high pressure (hypertension) or low blood pressure (hypotension).

In our view, *fluid pressure* is simply a physical property of some portion of fluid (in this case, blood), and the property can be measured without regard to whether the value indicates pathology or normal physiology. For the HPO, the assumption is that some human decision-maker has labeled some pressure measurement as abnormally high or abnormally low, and the ontology provides a consistent label for this abnormal phenotype.

In fact, an abnormality implies a comparison between two values of some physical properties. These could be measurements of that property over time, or between two instances of the physical object that bear that property. For example, when comparing cells, one can say that diseased cells have a smaller size. Thus, for phenotyping, one should characterize both the physical property and the value that is measured, so that one can make comparisons over time or against "normal". These ideas are captured by the Phenotype and Trait Ontology (PATO; Gkoutos et al., 2005), a resource that lists the "qualities" associated with a wide variety of phenotypes. A physical property is one type of quality; thus, PATO includes terms such as "increased mass", "decreased temperature", or "constricted". Notice that each also includes a comparative term, where the property measured has increased or decreased in comparison to some other state.

The word *phenotype* can have a very broad definition. In addition to the more narrow notion of an abnormal trait or property, phenotype can be defined as a larger-scale disease or syndrome. For example, the set of physical characteristics and behaviors that define diabetes or the cellular-level behaviors that define multiple sclerosis. In both cases, there is no simple genetic connection, and in both cases, there is no single attribute that can be measured. Instead, these sorts of phenotypes can be defined (with levels of uncertainty) by a collection of clinical attributes from the electronic medical record, including laboratory results, related diagnosis codes, and prescription data. There are a host of standards (sometimes ontologies) associated with electronic medical records. For example, the International Codes for Disease is used primarily for diagnoses, especially for billing purposes. The Systematized Nomenclature for Medicine is a separate, richer ontology for diagnoses, and there are also terminologies for laboratory results and medications.

However, this clinical notion of phenotype is not currently of value for our goals. Our aims are to obtain a mechanistic, biology-based understanding of phenotypes such as diabetes (as described in Chapter 1). In contrast, health record information contains very high-level patient characteristics of the disease. Even laboratory values, such as a higher or lower glucose concentration in the blood, are not directly connected to the pathophysiological processes that result in those abnormal glucose levels. Unfortunately, we do not even have a good language for describing these processes, at least not at the level of details that researchers would need to develop predictive models. This is one goal of the *Ontology of Physics for Biology*.

VISUALIZING BIOLOGICAL PROCESSES

As soon as researchers began studying biological processes, they began to draw these. For many biological pathways, processes can be envisioned with a schema of nodes and arrows, where the arrows represent the processes, and the nodes would be physical entities that participate in those processes. For example, Figure 2.2 shows some of the subcellular metabolic pathways involved in diabetes; Figure 2.1 shows the KEGG node-and-arc diagram for part of the mTOR signaling pathway.

Although intuitively appealing for human eyes, neither sketched drawings nor KEGG diagrams are standardized sufficiently for computer storage and use. There have been a number of attempts to formalize the visualization of biological processes, especially those that can be described primarily as a series of biochemical reactions. Figure 2.2 shows the use of the Systems Biology Graphical Notation (SBGN; Novère et al., 2009) to visualize an initial reaction for glycolysis.

Formally, such node-and-arc representations are directed graphs. Unfortunately, visualizations of large networks can quickly become too large and complex to be of much value. For example, Roche Pharmaceuticals have produced large wall charts that aimed to capture all cellular metabolic

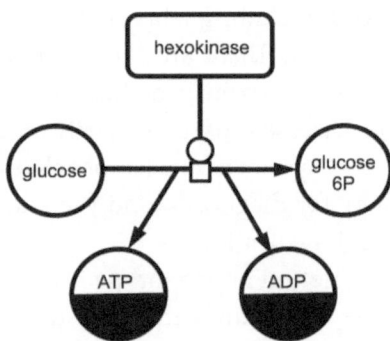

FIGURE 2.2 A diagram using the SBGN standard to show an initial biochemical reaction of the glycolysis pathway.

pathways. These physical diagrams on charts are so large and dense that, unsurprisingly, they are more recently available in an interactive format where users can zoom in and out to focus on regions and pathways of interest.

The challenge of how best to visualize complex and large directed graphs is an open research problem in computer science, with many potential layout schemes and interaction modalities. Cytoscape is a good example of a network visualization tool, one that is custom-designed for biomedical and genomic applications (Saito et al., 2012; Shannon, 2003). As one would hope, this tool includes the ability to read models described in SBML and display the relationship among these biochemical reactions as a network. Similarly, it can also view pathway information represented in the BioPAX format. However, these visualizations are auto-generated, and with large networks, the result is not pleasing nor useful. Instead of a structured diagram with recognizable motifs (such as in the hand-drawn Roche wall charts), the result can become a "hairball" diagram, where everything looks interconnected to everything.

Of course, Cytoscape and other diagramming technologies aim to avoid this problem, but a general-purpose solution is elusive, especially for large, well-connected networks. For the purposes of this book, we simply wish to alert the reader to this challenge; solving network visualization is beyond our scope. Instead, our focus is more fundamental; how can one specify the knowledge needed to understand biological processes? If we could create a catalog of such processes, could we assemble these to understand larger-scale biological systems, and pathologies such as diabetes or hypertension?

PUTTING IT TOGETHER: THE PHYSIOME VISION

In the early 2000s, biomedical engineers realized that if a cataloging of genetic sequences was of value, as the Human Genome project was demonstrating, then a similar and equally comprehensive cataloging of the processes that make up physiology would be of even greater value. This idea was given the name of "The Physiome" (Bassingthwaighte, 2000; Hunter & Borg, 2003). The vision is that if we can specify the physiome, then we can understand the processes that lead to disease, and then design drugs and interventions that disrupt these processes.

This vision is predicated on the existence of exactly the resources that we have listed in this chapter: To build the physiome, one must first catalog the set of all physical entities that participate in biological processes, and then build up a catalog of biochemical reactions, signaling cascades, and finally, larger-scale processes such as blood circulation. Again, the vision is that only then can we mechanistically understand the pathophysiology of diseases such as hypertension – a disease characterized by measuring a physical property, pressure, in the circulating blood.

This idea was compelling enough to initiate a number of high-profile projects: For example, DARPA funded investigators to develop mechanistic, physiological models of injury for the Virtual Soldier Project. Similarly, the NIH funded a project to carry out modeling for a "Virtual Physiological Rat Project" (Daniel A Beard et al., 2012). And finally, the European Union has funded the Virtual Physiological Human Project for over a decade, providing resources for a common institute (VPHi, https://www.vph-institute.org/) for coordinating a variety of Physiome-related projects.

> The vision of the VPHi is to ensure that the Virtual Physiological Human is fully realised, universally adopted, and effectively used both in research and clinic.

These projects and efforts have certainly made progress: As an example already mentioned, the PMR and the BioModels repository are examples of efforts that built on the foundation of this Physiome vision. Further, the physiome vision explicitly includes the notion of multiscale modeling and modular modeling, where larger, more comprehensive models of disease and physiology are built up from smaller models that can be composed together in a plug-and-play manner.

However, models cannot interoperate or be composed together if they do not share common knowledge about the biology and physics that they are modeling. As a simple example, if two models do not share terminology for naming reaction participants (such as common protein or small molecule names), then their points of biological overlap cannot be readily recognized. Indeed, establishing these common terminologies is one of the main goals of the bioinformatics informatics resources we have enumerated here. Unfortunately, as we have described, whereas the set of resources for physical entities is fairly complete, process naming schemes and measurable physical properties are less well-defined. As later chapters describe, providing a better ontology for these processes and properties has been a motivation for developing the OPB. However, before describing this ontology, we must first discuss and provide some additional examples of *ontologies*.

CHAPTER 3

Biomedical Ontologies

I N THE PRIOR CHAPTER, we provided a catalog and description of the wide variety of readily accessible knowledge resources for biological knowledge. Many of these are essential for modeling the "stuff" of physiology, but our catalog also demonstrates some important limitations of these resources. First, as we have described, there are many more rich resources available about physical entities (e.g., anatomy, protein catalogs) than there are about physical processes (biochemical reactions, or physiology). Second, as this chapter describes further, these resources are available in a wide variety of formats and styles and do not interoperate well. By this, we mean that often there is no easy way to connect concepts and information from one resource with information from another. As we will describe, there are some notable exceptions, and there are certainly research efforts underway to make these resources interact well with each other.

The goal of interoperability is to allow researchers to query and retrieve information without needing to know where (or in what format) that information resides. This goal is part of the FAIR principles: That scientific results be "Findable, accessible, interoperable and reusable" (Wilkinson et al., 2016). Especially for interoperability and reusability, different resources must share some common framework or definitions for terms that may appear in both resources. For example, if they share a common definition for "protein" that is precise about the meanings of terms such as isoforms and enzymes, then they can more easily interoperate. We call this sort of definitional framework of terms an *ontology*.

DOI: 10.1201/9780429469961-3

ONTOLOGY – THINGS, RELATIONS, CLASSES, INSTANCES

Ontology, as a "theory of being," is a branch of philosophy dating from Aristotle. However, in the 1990s, computer scientists (especially those working in knowledge representation) appropriated the term to apply more specifically to the terms and relations that one should capture for some specific domain of knowledge. In Gruber's definition, "…an explicit specification of a conceptualization that defines a set of representational primitives with which to model a domain of knowledge or discourse" (Gruber, 1995). In this chapter, we aim to make this academic-sounding definition more concrete, by providing biomedical examples of simple terminologies, taxonomies, and full ontologies that provide rich details and definitions necessary to fully disambiguate meaning across different knowledge resources.

To illustrate some of these ideas, we start with a "toy" ontology of anatomy expressed by the old spiritual "Dem bones":

> …Ankle bone connected to the shin bone
> Shin bone connected to the knee bone
> Knee bone connected to the thigh bone
> Thigh bone connected to the hip bone…

The song describes the normal anatomy of a vertebrate leg by naming five kinds of bones and how the bones are connected. This ditty describes some *things* (the bones) and one *relationship* ("connected to"). Further, because it describes these bones for any vertebrate, it can be thought of as describing bone *classes*. That is, your shin bone and my shin bone are two different *instances* of the more generic class "shin bone", by which we might mean all vertebrate legs. This *class–instance* distinction is a fundamental aspect of ontology and applies to multiple forms of knowledge representation.

If we continue with this example, we can imagine the "knowledge" of the song being captured in successively richer representations. Rzhetsky and others have labeled these as terminologies, taxonomies, and ontologies (Arp et al., 2015; Rzhetsky & Evans, 2011):

- *terminology* – a *collection* of defined terms that are ordered, say, alphabetically, according to names or identifiers that are established by a standards committee.

  ankle bone

  hip bone

  knee bone

shin bone

thigh bone

- **taxonomy** – a **terminology** that is organized **hierarchically** as a set of parent–child or class/subclass relations based on term meanings.

  Bone

  Long bone

  Shin bone

  Thigh bone

  Short bone

  Ankle bone

  Irregular bone

  Hip bone

  Knee bone

- **ontology** – a **taxonomy** of classes that also include more definitional information, and **relations** that can be used to connect classes. Thus:

  Bone

  Long bone

  Shin bone *connects to* Knee bone

  Thigh bone *connects to* Hip bone

  Short bone

  Ankle bone *connects to* Shin bone

  Irregular bone

  Hip bone

  Knee bone *connects to* Thigh bone

These choices are ways that we might formally describe information about the normal anatomy of an organism that could be you, me, a dog, or a wombat. The wonderful thing about the power of human communication and reasoning is that we interpret the song with many implicit notions. As humans, we can be both amused and informed by the song and the knowledge that it captures. However, if we wish to create a serious representation of leg anatomy that a computer can understand and process, we need to be much more explicit about representing anatomical knowledge. One challenge is that human intelligence is so adept at glossing over implicit assumptions, it can be hard for us to imagine how difficult representational problems are for something as smart as a computer.

The choice of anatomy as our first example is not accidental: As we aim to capture and represent human physiology and pathology, it is essential that we first understand and catalog precisely the physical entities that participate in those processes: The biological anatomical structure of *Homo sapiens*.

AN EXAMPLE ONTOLOGY: THE FOUNDATIONAL MODEL OF ANATOMY

Since our aim in this book is to characterize physiology, it is intuitive that a careful characterization of *anatomy* is essential to this study. The Foundational Model of Anatomy, or FMA, was developed by Rosse et al. (Rosse, Mejino, et al., 1998; Rosse & Mejino, 2003) in response to the need to reorganize content in the Unified Medical Language System (Lindberg et al., 1993). The authors were motivated by the need for a knowledge resource with the "requisite *granularity*, *semantic types* and *relationships* for comprehensively and consistently representing anatomical concepts in *machine readable* form" (Rosse et al., 1995).

We present the FMA in some detail for two reasons. First, we use the FMA as an exemplar to introduce key principles and best practices for ontology, such as a clear class taxonomy, formal class definitions, and explicit class-to-class relations tailored for the domain. Second, we demonstrate how we use the anatomical content of the FMA to represent anatomical entities that are participants in the two diseases described in Chapter 1: The cardiovascular system and its impact on hypertension and the metabolic/endocrine system and its impact on diabetes. The FMA is constrained to only include physical entities; we postpone discussion of the processes and models of those processes to later in this book.

FMA is a pioneering ontology in the biomedical domain. Historically, one should keep in mind that the development of the FMA during the 1990s co-occurred with our understanding of the value of the philosophical idea of *ontology* in computer science, and it preceded some of the more recent development of methods for knowledge sharing and semantic web technologies. The FMA established key principles and practices for representing multiscale biological structures from whole organisms to biological macromolecules (Rosse, Shapiro, et al., 1998; Rosse & Mejino, 2003). These principles and methods have informed other ontologies, including the Ontology of Physics for Biology (OPB), the central topic of this book.

Consider, again, the bit of folk anatomy provided by "Dem Bones". The snippet we quoted mentions five kinds of bone – anatomical entities – and declares that pairs of bones are "connected" by a type of anatomical structural relation. In ontology terms, each of the "bones" represents a "class" of bones within which a bone of yours or a bone of mine is an "instance" of that class of bone. If we look at the FMA, we find that each of these five bone classes is formally represented by an FMA class:

"ankle bone" = FMA:*Talus*   (FMA id=9708)

"leg bone"   = FMA:*Tibia*   (FMA id=24476)

"knee bone" = FMA:*Patella*   (FMA id=24485)

"thigh bone" = FMA:*Femur*   (FMA id=9611)

"hip bone"   = FMA:*Hip bone* (FMA id=16585)

The song describes how these bones are related by a *structural relationship* – "connected to", and this relationship can also be found in the FMA. In addition, all classes in the FMA are also interrelated via superclass–subclass relationships. Finally, most physical entities in the FMA have *parts*. For example, the FMA:*Tibia* has parts FMA:*distal epiphysis of tibia* and FMA:*Proximal epiphysis of tibia* (the upper and lower rounded ends). These relationships are a key formal definitional portion of the FMA, and indeed, most well-principled ontologies.

A fundamental relationship among classes is the **class–subclass** relationship. This relationship says that an individual of the subclass is also, by definition, an individual that is also a member of the superclass. Thus, this relationship sorts entities by "kind". A bone is a different kind of thing than a muscle, an ankle bone is a kind of bone but is a different kind of bone than a leg bone. Such classifications are established and reinforced by strict definitions that distinguish subclasses from superclasses and sibling classes from each other.

"ankle bone" = FMA:*Bone   organ*>FMA:*Short   bone*>FMA:*Tarsal bone*>FMA:*Talus*

"leg bone"   = FMA:*Bone organ*>FMA:*Long bone*>FMA:*Tibia*

"knee bone" = FMA:*Bone organ*>FMA:*Short bone*>FMA:*Patella*

"thigh bone" = FMA:*Bone organ*>FMA:*Long bone*>FMA:*Femur*

"hip bone"   = FMA:*Bone organ*>FMA:*Irregular bone*>FMA:*Hip bone*

In formal ontologies, such as the FMA, each subclass–superclass relationship is described by definitions such as the ones below for the class FMA:*Femur*, its sibling class FMA:*Fibula* (not mentioned in Dem Bones!), and its superclass FMA:*Long bone*:

FMA:*Femur*: "A long bone, each instance of which articulates with some hip bone, patella and tibia."

FMA:*Fibula*: "A long bone, each instance of which articulates with some tibia and talus."

FMA:*Long bone*: "A Bone organ, each instance of which has as its parts some diaphysis, distal and proximal epiphyses, and bone marrow cavity."

Defining information in the superclass is *inherited* by all instances of all subclasses. Thus, any instance of a femur or a fibula, or of any long bone, must also have distal and proximal epiphyses (the rounded ends) as parts. The aim of such formal, but human-readable, definitions is to clearly and formally distinguish subclasses from its superclass (e.g., FMA:*Long bone*) in terms of specific features of the subclass. In this case, FMA:*Femur* is distinguished from other long bones by anatomical articulations to other specific bones (the hip bone, patella, and tibia).

As a second, more complex example, consider the anatomy of the heart. The heart is a large organ with a complex anatomy and a variety of physiological functions that depend on physical phenomena at structural scales ranging from membrane ion transport to bulk fluid flow through the cardiac chambers. The heart includes anatomical parts such as atria and ventricles, valves, walls of atria, and the myocytes that make up the cardiac muscle. These parts can be arranged in a ***part-of*** hierarchy, where each child is part-of its parent class. For example, a myocyte may be part of a particular cardiac muscle, which may be part of the wall of the left ventricle, which is part of the left ventricle, which is part of the heart.

This hierarchy is entirely different from the *is-a* hierarchy of bones described earlier: whereas defining characteristics in an is-a hierarchy are inherited by subclasses, this is not true for a partonomy hierarchy. As we will describe later, partonomies are complex and hard to define formally. As one example, one might think that subparts could be defined as smaller objects spatially contained within a larger object. However, such a definition would imply that the portion of blood **contained in** a ventricle of the heart is therefore a **part of** the heart. In contrast, one might demand that the **part-of** relationship only holds true if the entity is no longer the entity without its parts. Thus, a heart without any ventricles isn't really a heart; whereas a heart without blood could still be considered a heart. At the end of this chapter, we will briefly discuss mereotopology: the formal study of definitions of the **part-of** relationship.

ONTOLOGIES: WHY?

We have shown how the FMA ontology provides a principled and comprehensive organization and catalog of the physical entities that participate in biological processes. While this is valuable in its own right, it is worth describing a bit more generally why one would build an ontology. As we introduced in the beginning of the chapter, ontology development should help acheive the FAIR principles: The goal should be to make knowledge resources *findable, accessible, interoperable*, and *reusable*.

With the advent of both "high-throughput" genomics and biology, as well as the ease of publishing data and knowledge onto the web, there has been an explosion of biological knowledge resources, such as we described in the prior chapter. For these resources to succeed, the terminology used by these resources must be precise and unambiguous. If resources use the same term to mean different things, or different terms to mean the same thing, chaos would ensue (as in the tower of Babel). Although this seems like a simple idea, getting community consensus on the details for particular terms can be challenging. Further, the inflexibility of computers exacerbates the challenge: For example, humans, unlike computers, are capable of quickly negotiating and building mappings between minor differences or misspellings of terms.

Further, building mappings across different resources does not scale if the number of resources is large. For example, Figure 3.1 shows the

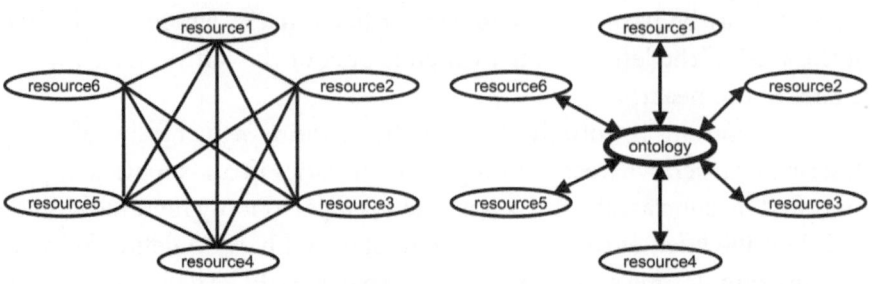

FIGURE 3.1   A common ontology allows resources to be interoperable with less effort.

possible relations among six resources; if we build all possible mappings, that becomes N(N-1)/2, or 15 mappings. Instead, if we have a common terminology, or common ontology such as the FMA, and all modeling resources make reference to that ontology, then it is much easier to make these resources interoperable.

An important example of this sort of ontology use is the Gene Ontology (GO) and gene sequencing experiments. Suppose a group of scientists working with *Drosophila* sequenced a possible gene and theorized that the gene plays a role in "transmembrane ion transport". To publish this result, they would *annotate* the gene sequence with this semantic term. In a high-throughput setting, there might be thousands of these putative genes and annotations. Then, if a separate group annotates a gene in the mouse as playing a role in "ion transmembrane transport", it will be hard for any automated system to notice this difference in phrasing.

The solution provided by the GO is to provide a taxonomy that all can use (Figure 3.1, right). The GO provides not only a canonical spelling, but also a hierarchy of terms, showing, for example, that "transmembrane transport" (GO:0055085) is a superclass of "ion transmembrane transport" (GO:0034220). In this way, systems can intelligently query across all resources that use annotations against the GO. For our purposes, we are interested in resources that describe physiology or pathology; therefore, in addition to the GO, we need ontologies for larger-scale anatomy, such as the FMA.

The vision of being able to comprehensively query web knowledge resources is not new. Indeed, the ability of common ontologies to support comprehensive querying was a key part of the Semantic Web vision (Berners-Lee et al., 2001; Hendler, 2001). In this vision, web pages and knowledge resources would be annotated not just with HTML tags for presentation/visualization, but also with semantic tags. These tags would include references to common terminologies or ontologies. Thus, users could leverage these semantic terms to more intelligently and comprehensively query across a wide range of web resources.

There are two requirements for this vision to succeed. First, a community must collaboratively develop an agreed-upon terminology, i.e., settling on the "spelling" and cataloging of unique terms. Second, the community should *define* these terms. As one can imagine, the latter is usually more challenging. Further, how should one specify the definition? As we have seen, the FMA uses descriptive, English-language definitions. Although this approach is much better than having no definitions, there are at least two significant problems with these natural language definitions. First, natural language is, by definition, ambiguous (unlike computer or mathematical languages). Second, as the number of terms grows large, creating these definitions in a consistent way does not scale.

As a solution, the semantic web vision offered the Web Ontology Language (OWL; Allemang & Hendler, 2011; Uschold, 2018) which includes precise, mathematical language for defining terms. For example, the GO includes the term "pentose-phosphate shunt", which can be viewed as part of metabolism. Like the FMA, the GO includes an English-language description of this term. However, in the OWL version of the GO (known as "go-plus", Blake et al., 2015; Mungall et al., 2014), the term also includes a formal, computable specification of a definition. As a rough translation, this definition states that

"pentose-phosphate-shunt" has-equivalent-class:

A "metabolic process" where

The property "has primary input" is filled with some "D-glucose 6-phosphate" and

The property "has input" is filled with some "NADP(+)"

The property "has output" is filled with some "NADPH"

The property "starts with" is filled with some "pentose-phosphate shunt, oxidative branch"

The property "ends with" is filled with some "pentose-phosphate shunt, non-oxidative branch"

This logic-based, formal definition allows for a wide variety of computational analyses over the GO. Note that all of the quoted terms above are also formally defined—in some cases by reference to external resources, such as ChEBI for NADPH. These types of definitions allow one to formulate, for example, a query that retrieves all "metabolic processes" that have a property "has output" filled with "D-glucose 6-phosphate" (and, thus, might be a potential predecessor to this process).

Further, as would seem logical, the property "has primary input" is a subclass of the property "has input". This means that a query asking about a particular value for "has input" would also return reactions with that value on the property "has primary input". Such behavior is an example of the system carrying out *inference* over the GO knowledge base. At an intuitive level, the system recognizes that when a user asks for information about "has input", the user also wants to know about "has primary input". For this knowledge base, this is a valid inference to make, because of the way these two properties are defined.

In general, the vision and hope is that formal definitions, accompanied by reasoning systems that produce appropriate inferences, will together make more "intelligent" and useful query systems. The OWL knowledge representation language is an essential component of this vision, as are inference systems that can "intelligently" answer sophisticated queries such as those described above. Many of the biomedical knowledge resources we described in the previous chapter use OWL, including ChEBI, the GO, Reactome, the FMA, and Uberon. However, some of these, including the FMA, primarily use the OWL syntax, without including the semantics of formal, computational definitions such as shown above for "pentose-phosphate-shunt" in GO-plus.

At this point, we offer some important caveats. First, this is not a book about knowledge representation or about the Semantic Web. Second, although the idea of the semantic web first arose 20 years ago, it has been, at best, only partially implemented. In our view, the challenge for full implementation of the vision comes from two problems. First, it is computationally expensive to carry out OWL inference over large knowledge bases (on the scale of the GO, for example), leading to slow or impractical query response times. Second, the semantic web vision depends on the development of large community-consensus ontologies. These are challenging to build. Indeed, one goal of this textbook is to communicate the challenges to building consensus around the semantics of biomedical entities, and hopefully to contribute to a better standard for biological processes and the mathematical models that capture hypotheses about those processes.

However, a clear success of the semantic web vision has been that the OWL language is recognized as a good choice for those aiming to support broad accessibility and clear semantics for ontology terms. As we present later, we have chosen OWL as the language for our OPB. Further, we demonstrate in Chapter 7, how we can use the inferencing capability of this language to enable the system to answer challenging questions about interconnected biological processes.

ONTOLOGY QUALITY

Given that we need to build an ontology for biological processes, how can we know if the ontology we build is a good one? How can we measure the quality or value of an ontology?

Early on, researchers have recognized that one should identify the purpose and scope of the proposed ontology *before* building the ontology (Fernandez et al., 1997; Noy & McGuinness, 2001). Thus, for example, the scope of ChEBI is to describe all "small" chemical compounds of biological interest (as its name implies) (Degtyarenko et al., 2008). The purpose of ChEBI and many ontologies is to provide unique, unambiguous names (IDs) for those chemical entities, as well as information about the structure and relationships among those entities. Although this informal specification of scope and purpose is not completely crisp (e.g., what constitutes "small"?), it provides at least a rough criterion for eliminating some types of content from the ontology.

The scope of the OPB is to define biological processes and physical properties that are useful when describing biology that involves the flow of thermodynamic energy (Cook et al., 2011). As we present in later chapters, all physical processes can be viewed through this physics lens as being a transfer of thermodynamic energy. We constrain the scope of the OPB to classical physics and systems dynamics, excluding quantum or relativistic physics (which are rarely invoked for biomedical processes). The goal of the OPB is to enumerate and provide details for these processes and the physical properties that change over time as these processes unfold. Where appropriate, we make direct reference to other ontologies that define physical entities, such as anatomic or chemical entities that participate in these processes. In Chapter 6, we will return to these goals and designs for the OPB.

Another way to think about the goals of an ontology is to ask: "What questions can this ontology answer?" These are sometimes called "competency questions", and they are a clear analog to the notion of software engineering requirements, which specify how a software artifact is expected to perform. For example, can an OPB-based analytical or modeling tool answer such questions as

- What processes does this chemical entity participate in?

- Is this process (and its participants) from the electrical, fluid, or chemical kinetic domains?

- What physical property (e.g., volume, chemical concentration, pressure) is modified by this process?

- What is (what are) the mediator(s) for this process? (e.g., enzymes for a chemical reaction)

Defining scope and goals are important initial steps, but once constructed, how do we evaluate how "good" our ontology is? Unfortunately, ontologies do not have a single attribute along which "quality" can be objectively measured. Instead, they are complicated artifacts which may serve a variety of different uses in different contexts. Because ontologies are central to the semantic web vision, there has been considerable research into how one might evaluate the "quality" of an ontology. This literature is quite large (e.g., see the International Semantic Web Conference proceedings); our goal here is only to sketch out a few basic ideas, and how (and if) our work developing the OPB aligns with some of these ideas.

Ontology Quality: Intrinsic Attributes

Some characteristics of ontologies can be measured or assessed without reference to anything outside of the ontology. For example, one can view an ontology as a graph (where classes are connected by relationships) and, therefore, one can easily measure attributes such as depth, branching factor, or attributes of connectedness (how many relationships connect classes how often). One can also look at the number of classes with definitions, or the number of classes with formal, mathematical definitions.

A challenge with these sorts of metrics is that they are descriptive, without providing much information about quality. Certainly, an ontology with a depth of one is probably not "high quality", but what depth (and what branching factor) is appropriate? An ontology with more relationships and more formal definitions is "richer" or contains more details than one that doesn't have these features, but is that always better? These questions are not answerable without considering external aspects such as competency questions and use-cases.

For ontologies that include formal mathematical definitions, or formal axioms about classes and/or relations, one can evaluate the ontology for inconsistencies. For example, an ontology could declare, via a formal axiom, that all biochemical reactions must have at least one reactant and at least one product. Then, if that ontology also specifies a type of biochemical reaction that has only a reactant, and no products, then we can state that the ontology is *inconsistent*. More so than the descriptive graph metrics, this measurement does have a clear consequence for quality – if an ontology is inconsistent, usually this impedes any question-answering system. Many inference systems will fail completely if the input ontology is inconsistent.

Ontology Quality: Extrinsic Attributes

In our view, extrinsic measurements are often a more useful measure of ontology quality. Unfortunately, they can sometimes be harder to measure. One simple extrinsic measure is how popular or well-used an ontology is (how many downloads, how many users, etc.). Although popularity is easier to measure, its connection to quality is not as clear. (The most popular kid isn't always the best choice for a job.)

Another extrinsic measure is to know how completely and how quickly an ontology can answer a battery of questions. The challenge with this measure is to know which questions to ask. As mentioned earlier, it is certainly important to understand the purpose and scope of the ontology so that the questions are "within bounds" of the scope.

Although many researchers have attempted to quantify ontology quality (Duque-Ramos et al., 2011; Fernandez et al., 1997; Guarino & Welty, 2002), ultimately, ontologies are a bit like other software artifacts—some things are measurable, but quality itself can be elusive, a bit like judging art. In fact, one effort at quantifying ontology quality is based on metrics for software quality (Duque-Ramos et al., 2011). This approach considers a wide-ranging holistic assessment, including reliability, operability, maintainability, transferability, and functional adequacy. Each of these has a series of sub-characteristics, many of which require extrinsic, subjective assessments by users. (For example, "operability" includes the notion of how easily a set of users can learn about the ontology.)

For modeling physiology, there are a number of interconnected knowledge topics that are important. As we reviewed in the prior chapter, there are online resources that cover many of these: resources for anatomy, biochemistry, cellular processes, and inventories of proteins and chemicals. Indeed, the number of available bio-ontologies is overwhelming—for example, BioPortal currently lists over 1,000 ontologies (as of 2023). Therefore, although we certainly wish to keep many of these broader ontology quality principles in mind, one of the most important ones is interoperability. That is, for physiological modeling, it is important for our ontologies to be able to refer to each other in a consistent manner.

UPPER-LEVEL ONTOLOGIES

The need for a variety of ontologies to interoperate and be consistent with each other is one of the main motivations for developing upper-level ontologies. The general idea of these ontologies is to capture the fundamental building blocks of reality – the superclasses that define all things, so that more specific domain ontologies can fit as subclasses of these broad, abstract entities. As this definition implies, the components of these upper-level ontologies are quite philosophical, aiming at defining things like "reality" and dividing up reality into discrete, well-defined classes.

As this is not a book about philosophy, we should be clear about why we care about such distinctions. The value of upper-level ontologies has to do with connecting different ontologies. For example, if two modelers use different ontologies to annotate and describe their computational models, how can we retrieve information from both and combine them appropriately? A non-scalable solution is to develop mappings that transform information from one ontology into the other. This is challenging because ontologies rarely have simple mappings (usually the terms' meanings overlap and intersect in complex ways). It is a non-scalable solution because there are too many different ontologies available. In contrast, if two modelers use different ontologies, but both ontologies subscribe to the same upper-level ontology, then the mappings problem, and connecting information across these ontologies becomes much simpler and manageable.

A key distinction, central to all upper-level ontologies, is between the physical "stuff" that exists in the world and the physical processes that occur as some stuff interacts with other stuff. Without clear definitions, there are potential confusions between physical entities and the processes in which they participate. In Chapter 2, we touched on this confusion when discussing E.C. numbers – do these refer to the physical enzymes, or to the enzymatic reactions that they catalyze? If this decision is left ambiguous, then information cannot be combined from the E.C. resource with other resources, such as the Protein Ontology, in a consistent manner. For example, whereas one can reasonably ask, "what are the other participants?" for a particular E.C. number (a question about process), one cannot ask that question of a protein in the protein ontology (which only includes physical entities).

Basic Formal Ontology

The Basic Formal Ontology (BFO; Arp et al., 2015) carefully defines the distinction between physical entities and processes: BFO:Continuant versus BFO:Occurrent. **Continuants** are "entities that *continue* or *persist* through time". In contrast, **occurrents** are "entities that *occur* or *happen*" and these processes unfold in successive phases over time. If something is an occurrent, then it has temporal boundaries; in contrast, one can discuss the structure of a protein or of the heart (continuants) without any reference to time. If a set of ontologies all connect or make reference to the BFO, then confusions about how to combine, for example, E.C. numbers with the Protein Ontology would be minimized. (In this example, the Protein Ontology does refer to the BFO, but E.C. numbers do not.) Figure 3.2 shows a partial view of top-level BFO classes.

As one might imagine for philosophy, there are several competing upper-level ontologies to choose among. For the purposes of this book, the key differences among them are their approach toward **realism**. The BFO is founded with the explicit goal of only representing "entities in reality" and excludes more ill-defined "concepts" that might exist in our minds but that do not have a corresponding entity in reality. In direct contrast, the

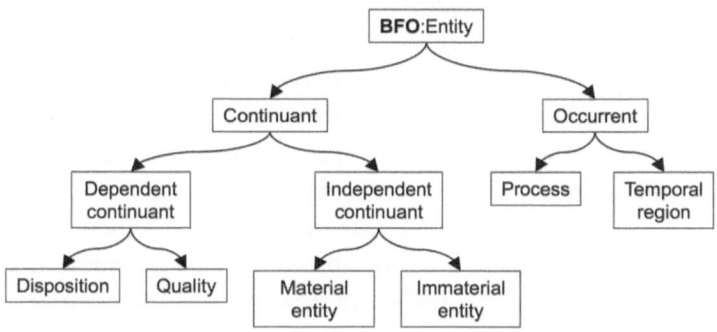

FIGURE 3.2 A pictorial view of the top-level classes of the BFO that map to the OPB representation of system dynamic theory as described in Chapter 6.

Descriptive Ontology for Linguistics and Cognitive Engineering (or DOLCE ontology; Borgo & Masolo, 2010) explicitly includes a pluralistic perspective that includes "concepts" that have linguistic representations. Likewise, the *General Formal Ontology* (Herre, 2010) not only includes "concepts", but also "levels of reality".

In general, our stance about such distinctions is governed by pragmatics: Do these upper-level ontologies support the sort of modeling we need for biosimulation models of physiology and pathology? Does the use of a common upper-level ontology promote interoperability among ontologies and terminologies? For the latter question, the BFO is notably successful among upper-level ontologies in that it is the foundation of the "Open Biological and Biomedical Ontology (OBO) Foundry", a large collection of ontologies that all refer to the BFO as their foundation (Smith et al., 2007, and see https://obofoundry.org/). Indeed, the BFO and the OBO collection have at least met the extrinsic quality measure of popularity, in that significant numbers of users work within this framework, and there are at least anecdotal examples of interoperability among OBO ontologies.

However, for the pragmatic question of meeting our representational needs, in a few places, we find the BFO framework lacking. In particular, we find that the realism perspective conflicts with our need to represent mathematical concepts. As we will describe in detail later (Chapter 6), biosimulation models often work with variables that participate in mathematical equations meant to represent processes such as reactions or fluid flows. Further, these equations are constrained by fundamental mathematical and physical laws such as Ohm's law, which can be applied to fluid flows such as blood in the aorta. An important and well-used concept in mathematics and in biological modeling is calculus: differentials and integrals. For biological modeling, it is often important to know that the volume of blood in a cavity or organ is the integral of the flow rate over time. Conversely, the temporal derivative of volume provides the flow rate.

Unfortunately, these pragmatic, mathematical concepts are exactly that: *concepts,* rather than entities in reality. For example, any definition of calculus relies on infinitesimally small changes, and therefore concepts like acceleration as measured for an infinitely small temporal interval are challenging to represent in a realist framework. A similar problem occurs in thermodynamics, where **energy** is only defined by computations on other physical properties and is not a directly observable entity in reality (Cook et al., 2019).

This limitation aside, we certainly do ascribe to the great majority of the organization and classes in BFO and other upper-level ontologies. As the example with EC numbers shows, it is critical to distinguish between continuants and occurrents – and this distinction is made in all three of the cited upper-level ontologies. We also recognize the importance of interoperability and the success of the OBO community in using the BFO as a common grounding for these ontologies.

As we described in the previous chapter, we are interested in the physical entities (continuants), processes (occurrents), and physical properties that are relevant to the study of physiology and pathophysiology. Properties that can be measured, such as mass, charge, and pressure, are central to our understanding of biological processes. Although the BFO does include physical properties as a type of "Quality", as we present in Chapter 6, we adopt a simpler hierarchy of top-level classes that correspond exactly to the three ideas of physical entities, processes, and physical properties.

Relations Ontology (RO)

Relations, such as *part-of* are the topic of the RO, which includes formal definitions of these sorts of relations (Smith et al., 2005). Precisely defining these relationships between entities is challenging. For example, an important consideration for *part-of* is that a sub-part should occupy part of the same spatial region as its parent. This is considered necessary, but not sufficient: recall the example of the blood in the ventricle—this blood is not usually considered "part of" the heart. Another complexity is that both occurrents and continuants can have parts. Thus, the process of making cookies can have a sub-part of "whipping eggs"; the longer process has a sub-process that is shorter and contained within the parent process's time boundaries. Although part-of can be a relational quality for both temporal and spatial objects, an important constraint is that processes cannot have spatial subparts, nor can continuants have temporal subparts. These issues around defining partonomy are referred to as "mereotopology".

As a second example, the RO also includes definitions for *located_in*, as in the "brain located_in head", and *contained_in* as in "lung contained_in thoracic cavity". The former relation is defined by referring to the regions that the object occupies and then noting that one of these regions must be part_of the other region. The *contained_in* relation is about location, but NOT parthood. Thus, it only holds between material and immaterial objects, such as anatomical cavities and lumens.

Ontologies of anatomy (such as the FMA) include many examples of these sorts of relationships, most notably *part-of* relations, as well as others, e.g., "*attaches-to*", which might relate muscle tissue to tendons to bones. For these relations, the RO provides a good resource for formal definitions.

OPB – THE QUANTITIES AND DEPENDENCIES OF CLASSICAL PHYSICS

In Chapter 6, we will introduce the OPB that extends available ontologies to represent, first, physics-based observable properties of continuants and occurrents and, second, the mathematical and logical laws by which values of such properties depend upon one another. In the next two chapters, we introduce and describe, first, how biophysical modeling is used for physics-based, biological systems analysis, and next, introduce engineering systems analysis as a theoretical and computational framework for analyzing, explaining, and simulating the behaviors of biophysical systems in health and disease.

CHAPTER 4

Biophysical Systems Analysis

W E HAVE INTRODUCED AND reviewed a range of biological databases, terminologies, ontologies, and other formal, searchable knowledge resources about biological *structural entities* and the *physical processes* in which they participate. Computational support for such "categorical" resources of discrete and qualitative knowledge have evolved, largely in the past few decades, and are now important tools for representing and reasoning about biological structures, processes, and qualities. However, such *categorical* resources have evolved largely independently of *quantitative, physics-based analytical methods* of physiology, biophysics, and bioengineering that trace their roots to Newton and Leibniz in the late 18th century and to Maxwell and others in the early 19th century.

Physiologists, biophysicists, and bioengineers are concerned with the observable physical properties of physical objects and the physical processes in which they participate. A heart is a material thing that contracts and expels blood in a process that involves the coordinated activity of its parts. A cell of one type may transform into a cell of a different type. A hormone molecule is released from one cell to control the metabolism of a cell of another type. These are *processes* that engage particular participating physical entities and occur over a span of time during which their participants are changed. It is the task of physiologists, biophysicists, and bioengineers to identify the participating structures and offer testable,

DOI: 10.1201/9780429469961-4

physical hypotheses about what happens during the process behaviors and offer cause–effect explanations for their occurrences.

For simple physical systems, processes may be sufficiently described and tested with hand-waving and simple sketches to represent mechanisms and predicted measurements. For the most part, however, physiological processes of biomedical interest are not so simple and require careful, objective articulation and rigorous empirical testing to be evaluated. In some cases, a simple hypothesis may predict an increase in some physical property value (e.g., a force or speed) when a decrease is actually observed.

Modeling "reality" has been a concern of philosophers upon which we will draw sparingly as our concerns are more prosaic and practical in scope. We use and advocate an engineering system dynamical approach that demands rigor and clear thinking but only insofar as they benefit better understanding and lead to practical solutions to bioengineering, biomedical, and bioscientific questions. Our observation is that physics-based models, whether intuitive "thought" models or solidly based mathematical models, help us think about and predict how the physical world works. In fact, we would argue that there are two benefits of modeling. First, that models can make explicit predictions of outcomes of physical processes in the biomedical world. Second, and possibly more significant for biomedical science, is that drafting, reviewing, and sharing a model demands and enforces a particular physical rigor and clarity of expression.

In the following sections, we will distinguish various "ways of knowing" and the roles played by various "models" in the biomedical research enterprise. We can distinguish some fundamental ways by which models differ within a domain and according to their intended use.

"MODEL"?

Our first task is to discuss the many meanings of the term "model" in the biomedical sciences. What are they? Do they come in different flavors? How good are they? In the following, we focus on various kinds of models as tools for solving problems in biomedical science and practice. Models are used to articulate and share our understanding of how the world may work in a way that is useful for explaining, predicting, or manipulating real things and phenomena. We take a very pragmatic view that is based on our own experience and is the basis for developing and deploying tools of use for the pragmatic solution of bioscientific problems. We find it useful to distinguish three broad approaches to modeling biomedical and biophysical systems such as *mapping, mimicry,* and *mechanism.*

For our purposes, a model is a *formalized expression of a hypothesis* that identifies a biophysical system's physical composition in terms of physical components: What are the participating entities? How are the entities related? And, in what processes do they participate? Modeling as hypothesis expression is an essential part of the scientific method and a key tool for managing the complexity of biophysical systems. Starting with simple pencil sketches of a system and its processes is a valuable mediator of scientific discussion and thought. Beyond that, mutual understanding of a model is strengthened by informal rendering in text or graphics but is much further strengthened by formal computational representation and analysis.

As a thought experiment, consider specifying and analyzing an *uber-model* that is comprehensive in all multidomain detail at all spatial levels. Thus, a model, to be useful and informative, must be drafted as a purpose-built abstraction that is intended to answer specific questions in the same manner as an experiment must focus on specific questions with a limited range of outcomes and specific interpretations.

Memory – Recall as Prediction

What we will call "mapping models" comprise a set of computational methods by which a set of empirical observations on a system are mapped to a set of predictions about future physical states of the observed system. If a physical system has, at a certain time, a set of certain physical property values, then one can map to a different set of values after some period of time. Its simplest form consists of observing, storing, and organizing data about a system and its components, and their properties that are mapped to a set of outcomes by "data-driven" methods based on conventional statistics and machine learning algorithms.

For example, medical practitioners observe and document a variety of clinical symptoms (e.g., chest pain, palpitations) and corresponding objective measures (e.g., heart rate (HR), blood pressure (BP)) that are recognized from prior mapping work of cardiologists and scientists to be highly related to a particular cardiological pathophysiology. This can be seen as a pattern-recognition task whereby certain clinical observations trigger certain diagnoses and, thus, therapeutic interventions. Physiological knowledge and clinical practices of this sort constitute medical school curricula and the medical practices of their graduates and take the form of recognizing patterns of symptoms and signs (i.e., test results) and opting for a therapeutic approach according to clinical guidelines and training. In its most basic form, this is a task of pattern-recognition/response applications.

These are the lessons taught during clinical training. Practitioners look for particular observable *signs* (objective measures) and listen for the patient's reports of *symptoms* (subjective reports), then formulate possible diagnoses according to currently accepted diagnostic and disease classifications. Such classifications may imply particular disease mechanisms (e.g., hyperglycemia due to a lack of insulin secretion) and suggest possible interventions and possible outcomes; e.g., pancreatic beta cells are dying and the patient requires insulin injections.

Model-free predictions can be based simply on statistical fits to observations without specifying physical mechanisms by which the behaviors are related to system inputs or to each other. For example, there are a variety of statistical methods to make outcome predictions without specifying or hypothesizing a causal model. For example, basic *linear regression* of a response value versus a stimulus value can usefully predict output values as proportional to input values without specifying a causal model. For dynamical systems, more elaborate statistical models can be used to fit observed responses and used to predict dynamic outcomes, again, without hypothesizing an underlying dynamical system. *Fourier analysis* decomposes a time series of data into a weighted sum of sine and cosine functions and then, by *Fourier synthesis*, can recompose a family of derived responses as outcome predictions by differentially weighting the sine and cosine functions. A more elaborate statistical modeling method is the so-called *Kalman filter* that iteratively weights the predictions of several predictive models – some of which may be mechanistic models – to find the optimum blend of model predictions that best predict the performance of real systems.

Artificial Neural Nets

Artificial neural networks have been developed since the 1960s by computer scientists who took their inspiration from basic signal processing in the nervous system (Krogh, 2008). Such networks consist of three layers of "neurons" (input, hidden, and output) each of which, as analogs to biological neurons, receive positive and negative "synaptic" inputs that are able to trigger neurons in another layer to "fire" when sufficiently stimulated. Neurons in the input layer are connected to those of the hidden layer which, in turn, are connected to those of the output layer. Input neurons can be activated in particular patterns which activate (or inhibit) hidden-layer neurons which then activate (or inhibit) output-layer neurons. How the input pattern propagates through the layers depends on specifying the "synaptic strength" to activate or inhibit neurons in a layer. Thus, an activation pattern of the input layer gets transmitted to the hidden layer and onto the output layer in a manner that can change according to the pattern of positive versus negative synaptic strengths.

This basic paradigm has been elaborated and adapted to a broad variety of tasks from basic image recognition such as needed for reading handwritten postal addresses to mapping gene expression patterns onto clinical outcome data. Computational investigators, recognizing the capabilities of neural nets, sought to apply them to recognition tasks such as reading and deciphering handwriting as needed to automate mail delivery systems. Computational neural nets have also been elaborated and generalized to serve as the computational basis of "deep learning" methods, which have been applied in the biosciences to map recollected past behaviors to likely outcomes. As such, these methods are examples of empirical memory models that occur in different forms with varying levels of rigor. However powerful as predictors, they offer little functional or mechanistic insight into how systems actually operate and should be considered "black-box" methods.

Much of clinical medicine training and practice is based on memory models by which practitioners are schooled in recognizing and memorizing cardinal features of physiological or pathological conditions in order to direct therapeutic decisions and predict physiological and clinical outcomes. "Programming" such a memory model is, however, a time-consuming task requiring years of model specifications – as clinical rules – and the clinical training to recognize patterns and apply the rules. The main task of medical education is to teach medical lore to physicians and practitioners and provide an empirical bottom line for medical decisions. Hence, the emphasis is placed on "evidence-based" medical practice; even if there are explanatory theories, their conclusions and implications must bow in the face of empirical evidence that contradicts theoretical predictions.

For example, a patient presents to a clinic with a blood sugar level that is twice normal. The clinic staff must marshal their combined clinical experience to predict the likelihood of various outcomes and therapeutic actions, such as: (1) the patient has recently eaten and absorbed four Big Macs that produced a marked, albeit transient, period of hyperglycemia, (2) the patient is an undiagnosed diabetic who will require further workup to manage his metabolic derangement, or (3) the patient is a diagnosed Type 1, insulin-dependent diabetic whose has run out of insulin and is headed for diabetic ketoacidosis, a frequently fatal outcome. Each such scenario defines a *current* state based on patient history (e.g., patient is normal, an undiagnosed diabetic, or a Type 1 diabetic), physical findings, and laboratory data that predict outcome estimates – "transition probabilities" – that a patient in a certain clinical state will transition to some other state.

At the other end of the temporal–spatial scale are state-transition models of molecular state changes. From college chemistry, one may recollect that a molecule of the sugar glucose can transition between two conformational states – "chair" and "boat" – and will do so with two transition probabilities: chair-to-boat and boat-to-chair. Large protein molecules have vastly more conformational states ("degrees of freedom") with state-transitional probabilities by which they change shape. Metabolic enzymes such as glucose-6-phosphatase can be in an enzymatically active or inactive conformational state depending on the metabolic demands of the cell. A potassium-conducting, adenosine triphosphate (ATP)-sensitive potassium ion channel (KATP) is a complex membrane ion channel that is composed of eight key protein subunits (four each of an ion-conducting subunit and an ATP/ADP binding regulatory subunit) that are each complex molecules with multiple conformations that depend on prevailing intracellular levels of ATP, adenosine diphosphate (ADP), Mg++, pH, etc. As a consequence of such complex mechanisms, a macromolecular complex such as a KATP channel has thousands and a correspondingly complex state-transition matrix.

What is common between clinical decision models and molecular state-transition models is that both are purely empirical. System states are defined by empirical observations, while the transition rates are expressed as statistical likelihoods. Such likelihood models are implicit in the clinical training and rules-of-thumb used for clinical decision-making and more formalized in observational data used to construct molecular dynamics models. Such models are purely observational and statistical, and so represent results solely in terms of probabilities of outcomes without invoking or representing mechanistic explanations. For example, traditional clinical knowledge, training, and reasoning are based on such pattern-recognition/response models because of the complexity of clinical practice.

An advantage of these methods is that they are purely empirical assumptions and make no *a priori* assumptions or theoretical expectations – predictions are based solely on the statistics of the mappings of inputs to outputs. However, there are two weaknesses of such models for predicting system outcomes – the "trust me" issue. First, predictions may be accurate only for a particular "training" dataset but predictions may be sensitive to other inputs. Second, the results are merely statistical correlations of outputs to inputs with no insight or explanations for the predictions. Thus, predictions can implicate, but not distinguish, identify, or explain the causal relations such as: "A causes B", "B causes A", or "C causes both A and B". Memory models provide scant insight into the system dynamics and exist as "black boxes" that offer no physical interpretations that can be reassuring to investigators or that can be used as a basis of hypothesis testing. There are no physics-based explanations as to how a particular outcome occurs. There is no opportunity to identify how a system can be refactored or retuned to achieve a particular outcome. In contrast, we and others have argued for more open "white-box" modeling that allows for better mechanistic insights as to how inputs become outputs (Neal et al., 2014; Yang et al., 2019).

Mimicry – Modeling "As If"

Medical electromechanical devices have no biological parts but are built, programmed, and tuned to behave as if they were the native tissues to replace failing physiological organ functions. A ventricular-assist device is an electromechanical model of cardiac ventricle contractile forces to assist ailing hearts. Even a hearing aid mimics and assists failing hearing by electronically mimicking the amplifying and filtering functions of the inner ear. The artificial endocrine pancreas can be programmed to infuse insulin according to prevailing and changing blood glucose levels to control blood sugar in insulin-dependent diabetics according to the dose–response characteristics of the normal endocrine pancreas. Perhaps, the most extravagant instances of biophysical mimicry are electromechanical robots that are entirely artificial devices consisting of coupled electromechanical mimics of sensory, cognitive, and musculoskeletal functions.

PID Feedback Control Systems

A frequent component of such physiological mimics is an engineered implementation of a generic feedback controller known as a "PID" (proportional/integral/derivative) controller. Engineers routinely design, build, or buy off-the-shelf PID controllers as devices to monitor and control process variables using feedback signals (e.g., injection of insulin) based on the deviation of a controlled variable from some desired "setpoint" value (Figure 4.1).

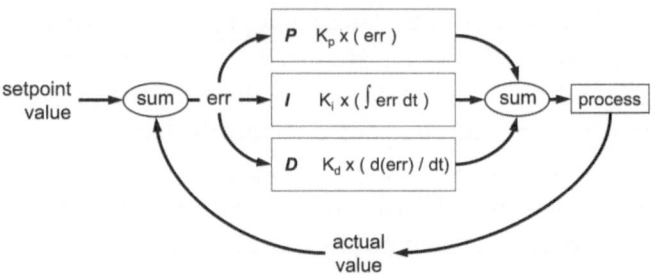

FIGURE 4.1 Diagram of a generic "PID" controller that calculates a feedback command signal that is proportional to an error signal that is the difference between a measured value and a desired "setpoint" value. The output signal is the sum of values that are proportional to (P), the temporal integral of (I), and the temporal derivative of (D) an "error" that is the difference between the actual measured value and the desired setpoint value.

As an example, a general-purpose "PID" feedback controller achieves blood glucose control by measuring a prevailing glucose level in the blood, calculating a deviation from an ideal setpoint level value, and injecting enough insulin to remove the excess glucose from the blood. In algebraic terms:

$$Gerr = Gactual - Gsetpoint$$

From this error value, the model determines a feedback control signal as the weighted sum of three separate mathematical functions of Gerr:

Proportional control: If the measured G deviates from a target G value (i.e., Gerr), secrete insulin as a feedback control at a rate (IRate) that is *proportional* (as Kp) to Gerr:

$$IRate = Kp * Gerr \qquad // \text{ proportional control}$$

Add *integral* control: if G deviation slowly creeps up (or down), increment (or decrement) IRate in proportion (as Ki) to the temporal integral of the G deviation.

$$IRate = Ki * \int Gerr \qquad // \text{ integral control}$$

Add *derivative* control: To damp out rapid fluctuations of G from food ingestion, add a component of insulin secretion rate that is proportional to the temporal derivative of the Gerr.

$$IRate = Kp * Gerr + Ki * \int Gerr + Kd * dGerr / dt \qquad // \text{ derivative control}$$

As shown in the generic PID schematic (Figure 4.1), "error" corresponds to Gerr which is used to calculate each of the three components (P, I, D) that are summed and fed as a feedback signal to the "Process" which, in this case, is the metabolic system by which insulin affects glucose level. Below, we discuss how such models may or may not be useful.

A second kind of dynamical model can reproduce physiological functions in very useful ways. For example, to correct some clinical conditions, it has been useful to create and program artificial devices to replace or bolster a failing bodily part. Medical mannequins are physical models that mimic one or more anatomical structures or pathophysiological functions and are used for training medical personnel in patient management methods such as for diagnostic exams, tracheal intubation, intravenous therapy, and patient transport. PID controllers can be applied and adapted to a wide range of control tasks precisely because they can be implemented and their PID parameters adjusted without any specific knowledge of the mechanisms that they are controlling. For example, a PID controller has been advocated as the control element of an automated insulin delivery pump for use by diabetics (Steil et al., 2006).

Such systems are particularly easy to implement in computer code, digital circuits, or analog electronics. However, implementing an electronic PID controller to replace or augment the function of failing insulin secretion or blood pressure regulation tells nothing about the normal physiological feedback mechanisms of islets or the hypothalamus. As such, a programmed PID controller is, at best, only a "reliable emulator" or "mimic" of a particular physiological function because none of the PID variables (e.g., Gsetpoint, Gerr) nor the separate PID signals correspond to measurable, or even definable, physical properties of the biological participants in physiological feedback.

What Sets the Setpoint?

Feedback control systems such as PID mechanisms require that some "normal" or "ideal" value be declared of the target value for the controlled variable. For example, one would specify a normal blood glucose level (e.g., 100 mg/dL) as the "setpoint" for an artificial pancreas used by a diabetic patient. In fact, diabetes researchers, when trying to understand how a real pancreas controls blood sugar levels often refer to a "glucose setpoint" and, thus, implicitly invoke the terminology of automatic control theory. Whereas setpoints can be defined and explicitly set for an artificial pancreas, there is no such "setpoint" that can be identified, measured, or inferred to exist in beta cells. There is no physical sample of glucose to which plasma values can be measured and compared.

Similarly, for physiological BP control, there is no compartment of fluid that maintains target fluid pressure to which BP may be compared. Rather, what appear to be "setpoints" are *emergent* properties of systems that behave "as if" they were a PID analog of a real physiological system (Beard et al., 2012). Although PID algorithms may mimic physiology, a physiological "model" must incorporate variables and functions that correspond to observable properties of physiological processes and their participants. However, the values of Gsetpoint, Gerr, and each of the parameters (Kp, Ki, Kd) have no physiological correlates and cannot be determined by physical measures.

One way of addressing this concern is to ask "what sets the setpoint?" which, in the PID architecture, is a value imposed by the modeler. In a fully functioning metabolic regulatory system, the answer to the question is far from clear. It is clear and apparent that if the pancreatic beta cell is the feedback regulator of plasma glucose level, then the researcher must (1) account for how such a glucose setpoint changes with age, body weight, stress, and health status and (2) observe, define, or propose a cellular entity or process that corresponds to a reference glucose level.

Emergent Properties of Complex Dynamical Systems

One is led, rather, to suppose that glucose setpoints and other critical or "target" physical property values are attributes of complete dynamical systems that are, in the final analysis, "emergent" properties of systems (Beard et al., 2012). An emergent property of a system is an observable physical property or attribute of a system whose occurrence is not inferable from even a deep understanding of the causal architecture of the system.

Engineers have a very thorough understanding of how PID systems work and they can predict with precision how changes to the system affect system behaviors and performance. Our assertion is that, however, useful is the "feedback setpoint" metaphor, it represents a rather superficial understanding of what "sets the setpoint" and how the setpoint changes in health and disease. Rather, from the perspective of emergence, feedback control and its apparent setpoint are observable properties of systems of sufficient complexity and operating at multiple spatiotemporal scales.

Pervasive examples of such emergent properties are provided by artificial neural networks that are, in one implementation, computational models inspired by the structure and processing power of biological nerve networks. Built as a set of interacting neurons that can activate or inhibit the firing of their neighbors according to the adjustable strengths of each such interaction. Neural nets can, therefore, be "trained" to perform certain computational tasks by repeatedly adjusting the interaction strengths to produce specific "output" patterns given particular "input" patterns. Early work, for example, showed that neural nets could be trained to recognize and identify hand-written letters and words despite differences in writing style.

The point we make is that although a given model such as a PID circuit or neural network may make useful and accurate predictions, *mimicry* offers little in the way of deep understanding of how real biological systems are constituted and the mechanisms by which they function.

Another approach is to mimic biophysical phenomena in one domain or spatial scale using analogous parts and functions in some other domain. Prior to the advent of general-purpose digital computers, a breed of analog computers were used to model all manner of dynamical systems from biological systems to aircraft structural dynamics and flight control systems.

Modeling on an analog computer consists of designing an electrical circuit such that its voltages and currents are analogous to and quantitatively represent properties of an observed or proposed system. As a physiological example, consider an analog model of a portion of the cardiovascular (CV) system that establishes electrical currents as analogous to blood flows and electrical voltage as analogous to BP that drives the blood flow. Dialing in a different voltage would represent changing BPs with consequent changes in electrical current representing blood flow.

Scale models are a kind of analog model in which structures and functions at one spatiotemporal scale are taken to be analogous to those at some other spatiotemporal scale. For example, the effect of heat on molecular diffusion could be modeled by subjecting a horizontal field of marbles (the "molecules") to a source of vibration (the "heat"). Or, the intercalation of DNA base-pairs was modeled at the human scale by the cardboard models of Watson and Crick. There are distinct advantages of analog modeling. First, they offer functional understanding and insight by direct and intuitive feedback if suitable functional analogs can be developed and defended.

A major disadvantage of physical analog models is the difficulty and expense of implementing systems in some interactive physical form. For example, nonlinear behaviors and physical properties are hallmarks of biological materials and phenomena. For example, biological elastic elements – tendons, muscles, ligaments – have distinctly nonlinear spring rates and viscoelastic and hysteresis behaviors that are difficult to reproduce in engineering materials. In the modern era, however, modern digital computers, graphic engines, and real-time internet access have provided analog modeling environments that are implemented and simulated on digital computer platforms.

Mechanism – Testing Physics-Based Hypotheses

"Memory" and "mimicry" approaches may be able to replicate and predict some behaviors of complex biophysical systems. However, apart from trial-and-error testing, neither approach offers insight into how the systems operate, nor how to identify opportunities to control its behavior. Physicians and biophysicists routinely rely on mechanistic understanding to make predictions based on physical understanding of what various empirical measures and observations mean about the future states of a patient or his disorders.

Bioengineers, physiologists, and biophysicists have, for years, borrowed extravagantly from the physical sciences to apply theoretical solutions to problems that had, or have, been identified and defined by the clinical sciences. In this section, we describe some, but not all, such approaches. At times, models in more than one paradigm have been combined into hybrid models that leverage capabilities of two or more paradigms.

Mechanisms as Causal Networks

System dynamic models are computational abstractions of real mechanisms based on the laws of classical physics. Depending on the needs of the research problem, models may be as simple as modeling the activity of a single chemical reaction or of blood flow through a single vessel. Or, models may be far more comprehensive and encompass complex sets of processes such as those controlling BP. The value of causal maps is that they use icons to represent various physical entities (e.g., molecules, cell parts) linked by various types of arrows that represent the processes by which they affect each other. As will be elaborated in Chapter 5, such network models illustrate the structure and functions of complex systems in a manner that can simply be read *qualitatively* as "increasing $Ca{+}{+}$ concentration increases glucagon release rate" or *quantitatively* using mathematical equations and variables to represent how glucagon release rate depends on intracellular calcium concentration.

A foundation of much of biophysical modeling is based on simple, first-order (i.e., linear) flow processes wherein the rate of flow is related linearly to the forces that drive the flow. Ohm's Law is the first of such relations whereby the rate of current (I, a flow rate) through a conductor is proportional to the voltage (V, a force) across the conductor, and inversely related to the resistance (R) of the conductor. Thus, $I = V/R$. Analogous algebraic "rate laws" apply to other kinds of phenomena such as a simple, irreversible chemical reaction flow rate (Q) is proportional to the concentration (C, also a force) times a reaction rate parameter (k); i.e., $Q = kC$.

However, more complex analyses may be required as exemplified by the following model that extends the PID framework discussed above.

Complex Networks: Adapting PID Models to CV Function

Generic PID modeling frameworks (e.g., Figure 4.1) can be adapted to emulate the behaviors of complex biophysical control systems; however, it may be difficult to identify the various internal variables with actual, observable physiological measures in the biological system. Rather, it is more useful and enlightening to take specific submodels as bases for modeling a system in terms of multiple, observable system variables such as HR, heart stroke volume (SV), vascular peripheral flow resistance (PR), and BP. A model of this sort can be tested, validated, and interpreted in terms of actual, observable system variables.

System-wide feedback loops for controlling have been the subjects of a large range of system dynamic simulations that model key physiological elements such as vascular blood flow, ventricular muscle mechanics, and neural control pathways. Other models focus only on a single blood flow or represent pathways using engineering control approaches. An engineering control circuit view (see below) of systemic BP control system based on the controls of SV, HR, and total peripheral resistance via the combined processing of the parasympathetic nervous system (PNS) and sympathetic nervous system (SNS) as controlled by the central nervous system. The control effort in this engineer's view of BP regulation is determined by an "error signal" that is the algebraic difference between a baroreceptor-measured actual pressure and a desired "setpoint" BP.

Models such as shown in Figure 4.2 are valuable because they can be tuned, tested, and validated against observable data points and, thus, are important for expressing, validating, and testing our understanding of complex, multidomain physiological systems.

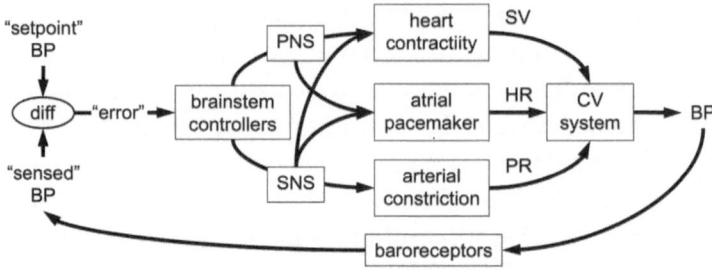

FIGURE 4.2 An engineering feedback control circuit view of systemic BP regulation whereby actual BP is sensed and compared to a setpoint value to yield an "error". The error value serves as a command signal that act via the PNS and the SNS to modulate key CV system parameters including ventricular SV, HR, and PR.

What do with the model?
Computational models are often dismissed skeptically as...
>**"You can *prove anything* with a model."**

However, this seriously misunderstands the scientific method and the role of models as formalized, computational statements of scientific hypotheses. Thus, the reality is even more pessimistic in that...
>**"You can *prove nothing* with a model."**

Hence, George Box, a scientific statistician, offered the following...
>**"All models are wrong, but *some may be useful."***

Source: WikiPedia: "George Box".

PHYSIOLOGICAL SYSTEMS ACROSS MULTIPLE SCALES

A most prominent and challenging feature of metabolic and CV systems is that their processes occur over multiple spatial scales – from atoms to organisms – and over broad temporal scales – from nanoseconds to creature lifetimes. Furthermore, organismic processes occur in multiple physical domains from biochemistry to fluid flow to neurophysiology. Here, we briefly introduce and describe a coarse stratification of physiological processes across multiple temporal and spatial scales.

Chemical Reaction Processes

Similarly, system maps of biochemical processes use the travel metaphor to map process pathways by nodes representing portions of chemical species (the cities) that participate in interlinked chemical reactions (the roads). Sigma Chemical, Inc., early on, offered a particularly comprehensive "map" of basic metabolic pathways that adorns the walls of many a biochemical laboratory to this day. The scope and complexity of such maps spawned a recognition of the need to computerize the vast and growing amounts of biochemical systems knowledge that was recognized in the early 1990s by workers at Kyoto University, Japan. The result is the ongoing KEGG project (the Kyoto Encyclopedia of Genes and Genomes; Kanehisa et al., 2023; Kanehisa & Goto, 2000) as an online database of chemical species as nodes linked by the reactions in which they participate (see Chapter 2).

Here, we discuss in more detail how various visual representations of biophysical pathway knowledge are useful for representing and reasoning about biochemical function. A major concern of such graphical languages is the extent to which the graphics unambiguously express the key structural and functional aspects of molecular species and processes. Consider a simple graphic representation of the phosphorylation of glucose (G) to yield glucose-6-phosphate (G6P) as the first step of glycolysis, the first reaction pathway in cellular metabolism is the conversion of glucose to glucose-6-phosphate (i.e., G → G6P) as a first step by which biochemical energy is extracted in the glycolysis pathway of intermediary metabolism.

Because this reaction is fundamental to our understanding of cellular metabolism, it has been rendered in innumerable diagrams using a variety of graphical conventions ranging from informal to formal. Some, as in the diagram in Figure 4.3, represent details of the chemical transformation of glucose into glucose-6-phosphate that includes details such as (1) the specific chemical site to which the phosphate group (OPO) replaces a hydroxyl group (OH), (2) that the P group is donated by a molecule of ATP to leave one molecule of ADP, in a biochemical reaction mediated by a single molecule of the protein enzyme hexokinase with participation of a magnesium ion (Mg++).

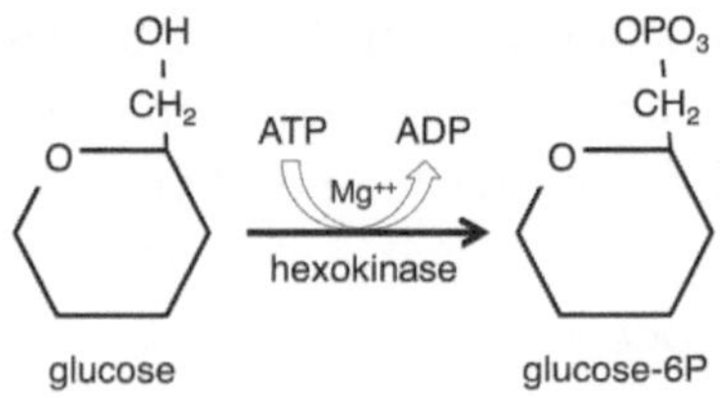

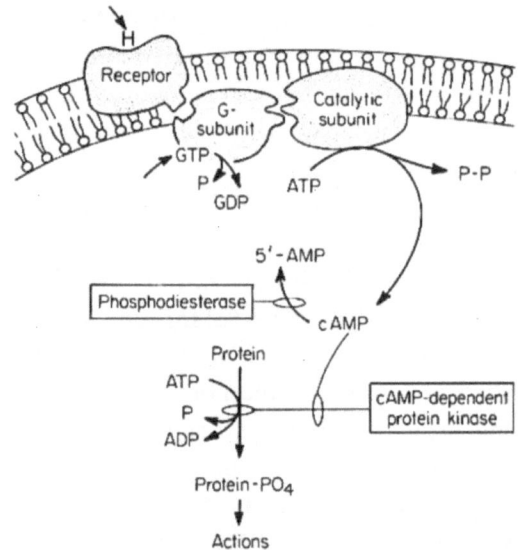

FIGURE 4.3 The key first step in the glycolytic pathway in which a molecule of glucose is phosphorylated to glucose 6-phosphate as facilitated by the enzyme hexokinase using ATP as a donor of the phosphate group.

FIGURE 4.4 Diagram of a generic cell-signaling pathway by which a hormone molecule, H, binds to a membrane surface receptor to activate a series of chemical reactions that (1) produce the intracellular "second messenger" cyclic-AMP (cAMP), which (2) activates a protein kinase enzyme that (3) phosphorylates specific proteins that (4) activate various downstream cell processes.

Molecular Signaling Pathways

Reaction processes such as the phosphorylation of glucose by hexokinase, as above, constitute a single step in extensive reaction networks such as the glycolytic pathway. However, similar cause–effect pathways are used to represent and reason about molecular signaling pathways. For example, changes of hormone concentration in the extracellular space get transduced into changes of constituent levels inside the cell without the hormone molecule actually being transported across the cell membrane as diagrammed in Figure 4.4.

Cellular Processes

Molecular processes occur within cells in the context of "organelles" that are subcellular structures such as the nucleus, mitochondria, cytoplasm, and the cell's plasma membrane. Membranes may contain embedded structural and transmembrane macromolecules that mediate transmembrane signaling and chemical transport. Figure 1.10 illustrates several basic cellular mechanisms within insulin-secreting beta cells such as mitochondria (left) that convert biochemical into ATP which closes ATP-sensitive K-channels which depolarizes the cell membrane, opens voltage-dependent calcium channels, and triggers the fusion of insulin secretory granules (upper right) to the membrane, and the release of insulin into the extracellular space (lower right).

Organ System Processes

In this very brief sampling, we have introduced notions that such systems are *multiscale* – from chemicals to organisms – and *multidomain* – involving various physiological and biophysical phenomena. For example, cells are parts of organs and cell processes constitute, correspondingly, the processes of organs and organ systems such as those that participate in blood glucose regulation in health and disease as in Figure 1.8 of Chapter 1. Such systems present real challenges for representation, modeling, and computational analysis that demand and begin with a focus on the physiological or pathophysiological problem. For diabetes, the focus might be glucose homeostasis (see Ajmera et al., 2013; Mari et al., 2020) – just how does this system work? What is the basis for its oscillatory behavior? What happens if a certain cell or organ is removed? These are the "what if" questions of a curious mind. Sometimes, there is a need to simply model the behavior of a biological system as an analog to be used as part of some diagnostic, therapeutic, or clinical training tool. Or, more rigorously, can we build a qualitative or quantitative model to actually test competing hypotheses for their adequacy for simulating and explaining empirical observations.

Whatever the motivation and goals, physiological modeling starts with an assessment of the scope and depth of available biophysical knowledge. This knowledge will help define the goals and scope of model development and testing.

REPRESENTING BIOPHYSICAL AND PHYSIOLOGICAL KNOWLEDGE

Pathway modeling provides *qualitative* information about the biochemical reactions and participants (proteins, small molecules, and cellular components) that make up biological processes. However, for complex reaction networks, a *quantitative* analysis of the processes is often essential for understanding behavior. Thus, modelers develop mathematical biosimulation models that provide mechanistic explanations for biological processes and behaviors.

These models can be written in standardized ways and, indeed, there are several languages specifically for biosimulation models. As discussed in Chapter 2, one of these, the Systems Biology Markup Language (Hucka et al., 2003), is the basis for a collection of more than 1,000 models available at the BioModels repository (https://www.ebi.ac.uk/biomodels/). A broader range of models can be found at the Physiome Model Repository (Yu et al., 2011), where the modeling language is CellML (Lloyd et al., 2004), a language that focuses more on expressing a model's mathematics. This collection contains hundreds of models ranging in scope from skeletal mechanics, to blood circulation, to electrophysiology, as well as metabolic and biochemical reactions.

In Chapter 2, we introduced several online resources for biochemical pathway knowledge that are the basis for cellular physiology, gene regulation, protein processing, metabolic pathways, and other components of the molecular biology of the cell (Alberts et al., 2007). Here, we discuss how this knowledge can be accessed and displayed to support intuitive reasoning over such networks to anticipate formal, computational methods that are introduced and discussed in Chapter 6.

Prior to the advent of cheap, fast, and powerful digital computer methods, bioengineers and biophysicists relied on a variety of graphical tools for representing the functional structure of the modeled hypothesis and for comparing model predictions (outcomes) with empirical observations. Diagrams have been used for millennia to represent and convey ideas and knowledge (Christianson, 2012) and then formalized to be interpretable by "diagrammatic reasoning" (Anderson et al., 2002). Back-of-the-envelope sketches propose explanations for how the parts of a system function. Musculoskeletal diagrams show how muscles move bones according to the structural constraints of joints and ligaments. Biochemical reaction pathway maps illustrate how various chemicals participate in a set of linked chemical reactions. Electrical circuit diagrams represent electrical components and the electrical signal-carrying pathways by which they interact. Transportation maps illustrate how geographical regions are linked by roads. Fortunately, humans seem to acquire, almost without instruction, how to reason through the functional implications of two key features of such diagrams.

For example, a road map represents the *structural topology* of a geographical region – Seattle is a city in King County which is part of the state of Washington. Detroit is a city in Michigan. Using one kind of topological reasoning one can determine that if one is in Seattle, one is also in Washington state and if one travels to Detroit then one is now in the state of Michigan. Furthermore, a road map can represent how regions are linked by roads to get from one place to another; to get from Seattle to Detroit, one can discover a pathway beginning on Interstate 90 with a turn onto Interstate 75 to arrive in Detroit.

Basic cell-phone apps can find this sort of travel itinerary using topological reasoning over database information about road–road connections in the interstate highway system. Such qualitative, topological reasoning is augmented by quantitative computations to estimate travel distances and times based on point-to-point distance and estimated car speeds for each road segment. Furthermore, our mapping apps are particularly useful because they can adjust their calculation based on local conditions and current travel speeds. This combination of qualitative, topological reasoning to discover pathways (as travel routes) and of quantitative, algebraic computation (of travel distances and times) readily generalizes for biophysical analysis.

We need to be clear in our terminology. For example, the "rate" of car travel and of chemical reaction processes can carry two meanings. One way to interpret the rate of travel is to track an individual traveler to determine the *rate of travel* of a given automobile on one trip from Seattle to Detroit; i.e., travel distance divided by trip duration. This view is an "agent-based" approach that tracks single automobiles, as individual "agents", on single trips from Seattle to Detroit during which each automobile has both an instantaneous rate (e.g., 40 mph traveling through Chicago) and an average rate (total trip distance divided by total travel time). The alternative interpretation of travel "rate" is the *rate of flow* of cars from Seattle to Detroit; how many cars arrive from Seattle on a given day in terms of cars per day. This interpretation and view is a "kinetic" or "flux" approach that we will develop, below, as "stock-flow" modeling.

CRAFT OF QUANTITATIVE MODELING OF MECHANISM

Our discussion of analytical tools used in biophysical modeling will assume a working knowledge of basic college mathematics including arithmetic, algebra, and both plane and solid geometry. This book is not the place for an in-depth discussion of the following analytical tools, which we include here simply as an introduction to the concerns addressed by the analytical modeling community but which are covered in depth in many other resources.

Mathematics – Differential and Integral Calculus

The key basis of calculus is the separate recognition by Newton and Leibniz that for a continuous value, y, that is a continuous function of another value, x (i.e., $y=f(x)$), one can imagine two critical mathematical relationships that are fundamental to the analysis of time-dependent dynamical systems as encountered in biophysics. First is their (Newton and Leibniz) recognition that the ratio of a tiny change of x, Δx, to a corresponding tiny change in y, Δy is, then, the *differential* or *slope* of a curve of $y=f(x)$ in the region within the span of Δx.

The assertion and insight of Newton and Leibniz was that the *differential*, $\Delta x/\Delta y$, would approach the curve's *derivative*, dy/dx, which is the slope of a *continuous* curve, $y=f(t)$, at a single point on the curve. This conclusion continues to trouble many mathematicians and philosophers (Strogatz, 2019), who argue that the value of a continuous function cannot, by definition, change at a single point. However, the convergence of a discrete *differential*, $\Delta x/\Delta y$ to a *derivative*, dy/dx, is the slope of a curve at a given value of y. This assertion is an essential step in the *differential* calculus as used for expressing temporal rates of change such as for *velocity* as the temporal rate of change of spatial position and *acceleration* as the temporal rate of change of velocity.

Variables vs. Parameters

Modelers draw a strict distinction between model "variables" and model "parameters". A parameter is a quantity that characterizes some aspect of a modeled system and is, generally, a constant whose value distinguishes one instance of a system from other instances. A variable is a quantity whose value changes during a modeled process. For example, the flow rate (F12) of blood through a vessel is proportional to the pressure gradient (P1–P2) from one end of the vessel to the other and is inversely proportional to a resistance parameter (R) that depends on vessel geometry and the viscosity of blood. Thus, $F=(P1-P2)/R$.

In this formulation, F, P1, and P2 are model variables whose values change with time during a process, whereas R is a parameter whose value may be taken to be a constant during a process. More specifically, R is a constitutive parameter (or sometimes, a coefficient) because its value depends on the spatial and material constitution of the experiment.

Specifically, the flow resistance depends on the length and cross-sectional area of the flow path and on the viscosity of the flowing fluid. Thus, this formulation is used routinely in engineering applications for the so-called Newtonian fluids, such as water, for which viscosity really has a fixed value across a broad range of conditions.

However, blood is not a Newtonian fluid in that its viscosity increases as the blood flow rate slows because slower flow allows more interactions between the blood's cellular and protein components. At slow flow rates, blood cells and protein have more opportunity to bump into and adhere to each other and so increase blood viscosity (this is the same effect seen as the stubbornness of catsup to flow slowly). For the same reason, blood viscosity is higher in smaller vessels, particularly in capillaries where there is much more opportunity for blood constituents to interact with vessel walls.

There are many ways that a computational biologist can quantitatively account for such effects to extend the scope and range of an analysis. For example, one might derive an equation that expresses such effects in terms of vessel geometry and fluid flow rates. Depending on the formulation and range of these effects, one may be able to derive a formulation such that "blood viscosity" is a variable that is expressed in terms of other constitutive parameters.

Getting the Best-Fit – Parameter Optimization and Sensitivity

A pervasive challenge of biophysical research and biomedical practice is managing and accounting for the inherent variability of biological entities and properties. For the most part, engineering technological products, motors, cars, machines, etc. are so precisely engineered and manufactured as to have very narrow variations of size, weight, strength, rates, etc. For most applications, engineers need to pay only cursory attention to statistical issues beyond calculating average values of property measurements. However, even simple biomedical measures can be fraught with random variations that confound even simple assertions and require statistical methods. One might want to simply determine the statistical mean and variance of a dataset or determine the *slope* and *intercept* of a line that provides the *best-fit* for correlating values of two measured variables. However, a dynamical model can have many parameters and many output variables, which means that *optimizing* the parameter values to get the *best-fit* for a mathematical model to a data set is a more difficult problem as discussed in Chapter 5.

What set of parameter values *optimizes* the model's fit to a data set? How *sensitively* do model predictions depend on changes in model parameter values? Whereas finding the best linear slope and intercept of a line

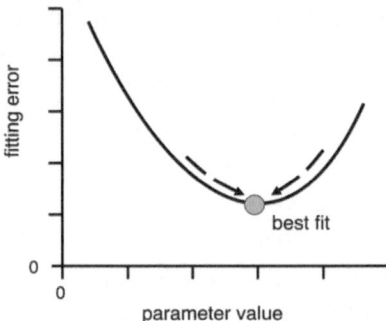

FIGURE 4.5   Illustration of a "gradient descent" algorithm to minimize the error (increasing on the vertical axis) of a model prediction by varying values of one or more model parameters.

to correlate is a straightforward computation, complex models with multiple internal variables and multiple parameters require more elaborate but analogous methods. The challenge is to establish a set of parameter values that minimize the error between model predictions and a target empirical data set through a process called "model optimization". Whereas statistical correlations can be computed directly from datasets, optimization of models having more than one or two parameters requires sophisticated iterative, error-minimization methods. For example, Figure 4.5 is a plot of fitting error on the vertical axis, as a function of the values of a model parameter value to identify the optimum value that minimizes the error between a model prediction and experimental observations.

We have so far taken an in-depth look at some of the nut-and-bolt methods of physics-based modeling. Now we turn to visualization methods for representing the overall structure of such models after which we consider in broader perspective the meaning of biophysical "model".

MODELING SCOPE AND SCALE

Chapter 3 introduced the fundamental ontological distinction between *continuants* and *processes*. In this chapter, we reiterate and expand these distinctions using examples that span multiple spatial and temporal scales across multiple physical domains. Chapter 6 will describe how we have formalized these ideas and encoded them in the Ontology of Physics for Biology (OPB). Within the scope of classical physics, physiologists, biophysicists, and bioengineers task themselves with explaining, understanding, and analyzing both normal and abnormal biological functions from the molecular to the organismal level. This broad scope is partially expressed in Figure 1.2, which illustrates the span of spatial scale ($\approx 10^{12}$) of the structural entities, temporal scale ($\approx 10^{12}$) of process durations, and the sheer number of different types of, for example, molecules, cells, or organs.

Consider the fluid dynamics of the flow of blood, a distinctly non-Newtonian fluid, through blood capillaries barely the size of red blood cells themselves. Or, consider turbulent blood flow as it passes through dynamically opening and closing heart valves and into the root of the aorta. Consider the microsecond dynamics of ion channel opening and closing to regulate the flow of picoampere ionic currents that control the synaptic activity of neural transmission. Or, consider the profoundly intricate coupling of cell-signaling processes that integrate metabolic and neural inputs to regulate the secretion of insulin from beta cells. And finally, consider how the overall operation of an organism is the integrated expression of all such processes occurring within an individual organism in health and disease.

In our efforts to represent, in a computable form, the broad range of biomedically relevant physiological processes, we have strived to conform to and leverage biomedical knowledge representation standards as have been developed in the past 10 years and described. As we note below, however, although these ontologies offer starting frameworks, the scope of biophysical analysis demanded extensions such as ways to represent electrical charges and fields, thermodynamic energy, and entropy, and the broad range of physical laws and axioms. Here, we describe the scope and span of biophysics as represented in the OPB.

Space and Time Are Continuous and Unbounded

Quantum physicists assure us that both time and space are, at root, quantized and not smoothly continuous as has been assumed by classical physicists since Newton (Rovelli, 2018). However, the tiniest temporal intervals (e.g., nanosecond; 10^{-9} seconds) and spatial spans (e.g., nanometer, 10^{-9} m) that concern biophysicists are orders of magnitude larger than the corresponding quantal dimensions of Planck time (10^{-44} seconds) and Planck length (10^{-34} m) that are of interest to quantum physicists. This means that for all practical purposes, spatial and temporal spans such as the volume of a cell or a rate of blood flow can be treated as continuous functions.

Although space and time are usefully measured in discrete intervals – e.g., meters and seconds – classical physics is based on continuous measures of both time and space. Neither has an origin or end, and neither has a natural partition or increment. Each is a continuous scalar quantity quantified, typically using "t" to represent a span of time, and "x, y, z" as distances from defined origins along each of three orthogonal axes in a Cartesian spatial coordinate system. Both space and time are continuously differentiable and integrable using the mathematics of classical differential and integral calculus developed in the 18th century by Newton and Leibniz.

Object vs. Process Models

Material objects are composed of atoms, and each atom is composed of subatomic particles in various ratios. Consequently, the amount of matter in all material entities – i.e., composed of atoms – is the sum of the masses of all the subatomic particles comprising all of the atoms in a material entity. Molecules are composed of atoms that can be counted for stoichiometric calculations in chemistry, and for micro- and nanoscales involving a countable number of macromolecules such as the genes in a cell nucleus, or the bits and pieces of the cytoplasmic machinery (e.g., ribosomes), or the membrane ion channels that can occur in countable numbers per cell.

Depending on the biophysical system being studied, the data that are available, and the nature of the scientific questions to be posed and answered, some analytical approaches are better than others in terms of how continuants and processes are abstracted in terms of continuous vs. discrete mathematics (Table 4.1).

ODEs, by far the most prevalent biophysical modeling and analysis methods, are based on equating the temporal rates of processes (e.g., insulin secretory rate) with temporal rates of change of continuant properties (e.g., rate of change blood glucose) of discrete continuants (Ajmera et al., 2013).

Partial differential equations (PDEs) are used to model temporally continuous changes within continuous spatial distributions of, say, stress vectors within a solid muscle, or flow rate vectors within a portion of flowing fluid. Prime examples of the power of the PDE approach are apparent in simulations that merge PDE models of cardiomyocyte electrophysiology with models of ventricular contractile mechanics and with ventricular blood flow (Kerckhoffs et al., 2007; Nickerson et al., 2016).

TABLE 4.1   Chart of Analytical Approaches as Appropriate to Available Knowledge and Computational Tools

		Process model	
Object model	*Space* / *Time*	*Continuous*	*Discrete*
	Continuous	Partial DE	Logical inference
	Discrete	Ordinary DE	Agent-based

Agent-based algorithms are used to model the replication, destruction, and interactions of discrete, spatially distinct continuants such as single red blood cells as they move through a continuous capillary lumen (An et al., 2009; Bock & Gruninger, 2005; Sluka et al., 2014).

Logical inference models (e.g., Markov methods, Petri nets) assign discrete states to continuants (e.g., bound/unbound) that can change (e.g., bound → unbound) at discrete moments in time (Peleg et al., 2004; Wang et al., 2012).

Finally, one can develop *Multimodal approaches* that integrate submodels expressed in different modeling modalities such as models that integrate ordinary differential equation (ODE) models of cell metabolic processes with logical inference models of genomic signal processing (Karr et al., 2013).

Collections of Discrete Things

One of the more challenging aspects of biophysical analysis is the sheer numbers of entities and processes that must be represented as well-defined sets (or collections, or populations) of individuals. A set in mathematics is a collection of distinct objects as defined by some explicit criteria. A set is an object or entity in its own right so that one can have sets of sets.

As an example, consider the set of all bones in an individual human, or the set of ribosomes in a particular cell, or the set of genes in an organismal genome. If one were interested in the oxygen-carrying capacity of blood in a coronary artery, one could define a set of all the red blood cells in a coronary segment. If the cells are identical, then the total oxygen capacity equals the number of cells, N, times the oxygen content of the average single cell. Or, if not identical, then the amount of oxygen per cell is represented by the distribution of the number of cells in subsets of cells with a particular oxygen content.

Each such entity is enclosed by a spatial boundary. Three-dimensional entities are enclosed by 2-D surfaces; that are themselves bound by 1-D lines that are bounded by 0-D points. Accordingly, each 3-D solid is a *volume* property (V) that is a measure of the space within its boundary and, correspondingly, each surface has an *area* (A), each line has a *length* (L), and each point has a coordinate *location* (x,y,z) in a spatial reference frame.

KINDS OF PHYSICAL MEASURES

In prior chapters, we have described bioinformatic resources that name, catalog, and classify biophysical continuants and processes but these resources do not represent or explain how biophysical continuants change during these processes. Such analyses are concerned with the *observable attributes* of physical things – the *size* of a continuant or the *rate* of a process – rather than how such things are classified. Such observable measures are required to express and test biophysical hypotheses based on the quantitative physics of the continuants and processes. In the following, we will describe basic observable physical measures, their measurement units, as well as their variations in time and space.

Biophysical system dynamical analysis aims to account for observable values of physical properties (or "qualities" or "attributes", in some nomenclatures) as they change, or not, in real time and real space. As will be described formally in Chapter 6, we take a biophysical property as an attribute of a biological continuant or process that can be, in principle, observed and measured by physical means. This includes fluid pressure and flow rate, chemical concentration, and body height. It will not include more observational and clinical phenotypes such as "hyperactive", "feeling ill", "conspicuous", or "fertility" included in medical phenotype ontologies such as PATO (Phenotype And Trait Ontology; Gkoutos et al., 2005). Rather, we will consider those attributes that are observable by physical means that will be more formally defined and classified in the OPB described in Chapter 6. Before we focus on system dynamic analyses of biophysical systems, let us distinguish a variety of attributes of biophysical continuants and processes.

Physical Measures in Biophysics and Engineering

We discuss observable physical attributes as they are introduced, defined, and used for data collection and computation. Thus, we first rely on an intuitive understanding whereby "weight" corresponds to the "heft" of an object, or the extent a spring stretches to suspend an object. By instruction and usage, students of physics and engineering come to tacit, but loose, agreements as to such meanings. Whereas the meanings and calibrations of such properties become honed through usage, such an informal treatment has resulted in a tangle of informal property definitions and competing units of measure. We will focus here on those physically measurable properties of entities and processes that are represented by OPB for system dynamical analysis. In so doing, we will take a broader view, define and distinguish kinds of measurements and their respective units as they apply to biophysical entities and processes. This entails introducing the vocabulary of physical measures and concepts that engineers and physicists have identified such as distinguishing the mass of an entity from its mass density.

Biological Measures Present Quantification Issues

As critically useful and comprehensive as the SI and the unit ontologies have been for facilitating biophysical research data recording, archiving, and exchange, their utility is critically limited by three challenging aspects of biomedical units (Hemker & Beguin, 1993; Lehmann et al., 1996; Schadow et al., 1999).

First, clinical observation and medical practice have developed over centuries and have long grown out of clinical procedures that have preceded the development of measurement methods. Hence, we have (1) temperatures measured in degrees-Fahrenheit (°F) rather than degrees centigrade (°C), (2) BPs measured in millimeters-of-mercury (mmHg) rather than, say, a more SI-compliant dynes/cm^2, and (3) the plasma concentration of hemoglobin A1c in millimoles as a proxy for the average of the plasma glucose levels over the preceding week that are directly measurable in millimoles.

Second, like clinical medicine, other biomedical and biophysical domains of study confront a wide variety of anatomical structures, processes, and properties that have been named and defined according to domain-specific needs. For example, the domain of biological mass transport and chemical exchange has evolved a set of specialized terms to analyze, for example, vascular blood flow and transmembrane protein exchange (Bassingthwaighte et al., 1986). This terminology includes domain-specific properties and terms such as (1) hct=hematocrit, (2) k_F=filtration coefficient, and (3) L_{pD}=osmotic coefficient. Given the range of entities, processes, domains, and temporospatial scales, the field of biomedical research and knowledge representation is rife with terminology and annotation issues that present major problems for resolving unit mismatches in cross-disciplinary computations as described in Chapter 6.

Third, the results of many laboratory tests are expressed in units that are relative to some arbitrary standard as, for example, the "activity" of an enzyme in a clinical assay may be calibrated against the activity of a "standard" sample but whose actual enzymatic activity may be unknown or is unknowable.

Continuous vs. Discrete Measures

Many physical properties such as spatial or temporal location are inherently continuous in time and space; we can imagine that they can be measured and specified at arbitrarily small scales. The precision of such measures is necessarily limited by the precision of the measurement instrument or the numerical representation (e.g., 8-, 16-, or 24-bit). Other physical properties, such as the concentration of red cells in a small blood sample, are inherently discrete because they may ultimately represent the countable number of discrete entities (e.g., n = 200 cells) in a very small amount of blood or represent the "hematocrit" as a continuous measure for larger blood samples.

Measures of inherently continuous physical properties, such as time or spatial location, are necessarily discretized by both the precision (e.g., ±0.001 V) of the measurement instrument and the numerical precision (e.g., 16-bit vs. 32-bit) of the analysis or simulation algorithms. For example, a thermometer with a precision of ±0.1°C discretizes temperature measures such that both 37.12°C and 37.14°C may be read as 37.1°C.

Categorical Measures

Qualitative attributes such as the geometrical shape (e.g., cuboidal, spindle-shaped) of cells in a population may be useful for identifying and categorizing cell types. However, investigators must take pains to define inclusion/exclusion criteria in sufficient detail to assure that the categorization is reproducible by others viewing the same dataset. For example, the number of red blood cells in a sample or the number of ion channels in a cell membrane. Such measures must be defined for the common use of cell-sorting machinery whereby a population of individual cells can be automatically sorted by type according to the results of laser scans that distinguish, say, cell size or cell color. Such data and discriminations are critical for population kinetic models of observable changes using various if-then rule sets as in agent-based dynamic modeling (An et al., 2009; Blinov et al., 2017; Sluka et al., 2014).

Categorizing colors can be challenging. Fuzzy logic (Kosko, 1993) was once advanced as an approach to modeling and reasoning about categorial information where classificatory criteria may be ambiguous or uncertain. For example, various human observers may consistently classify and distinguish red and yellow things, but will differ in the classification of things that are edge cases that may be orangish, reddish orange, or yellowish red. The computations at the heart of fuzzy logic aim to account for the ambiguities of color value over the range of color values at the boundaries between red and yellow. Such approaches depend on simple linear interpolation coupled to various propositional logics. Whatever the analytical potential for fuzzy logical methods, they have yet to significantly penetrate the domain of biophysical modeling.

Population Measures

Certain quantities can be measured using discrete counts that are appropriate for tallying the number of electrical charges on a single molecule, the number of amino acids comprising a protein, or the number of genes in a cell, or the number of cells in a microscopic tissue sample. The number of atoms or molecules comprising a macroscopic object, although expressible as discrete counts, results in such huge numbers that the counts are best treated as continuous, real numbers. For example, the number of atoms in a mole of a chemical is Avogadro's constant (6.02214×10^{23}) that requires a decimal number having 23 significant digits. Such a large number far exceeds the capacity of available computational machinery and exceeds the precision of available measurement methods. Consequently, amounts of material are routinely expressed and computed using more modest levels of precision. Whereas the engineering and physical sciences may demand measures and computations with precision on the order of 1:10,000 (10^{-4}) or less, the biomedical sciences deal with such large, highly variable measures and phenomena as to routinely encounter variabilities of 10% or greater.

ATTRIBUTES OF PHYSICAL MEASURES

Precision vs. Accuracy

The *accuracy* of a measured or computed value is how close the value is to some *true* value. The *precision* of a value is the closeness of repeated measures or computations of the value. One can obtain a very accurate measure that is, say, within 1% of the "true" value but which is, however, not very precise because the measure is very poorly reproduced when determined repeatedly.

Number Forms

A real number is the value of a continuous quantity that can represent a distance along a continuous line such as the temperature of a material entity or the mass of a material object. The value may be an integer value (0, 2, 20) or have a fractional component (1.34, 4.3) and may include a multiplier that is a power of 10 (3.4E2 = 340). Table 4.2 is a brief overview of various number forms and notations used for biomedical data collection and model computations. Such distinctions are absolutely critical for proper computation.

TABLE 4.2   Scientific Number Forms and Notations

Real – may include fractional part
    base-10, decimal for quantification: 1, 2, 3,...101, 102,...
    may be in "E-notation" multipliers of powers of ten
        e.g., 124 = 1.24E10, 0.05 = 5E-2,
Integer – a real number that lacks fractional part; may be + or −
    base-10, decimal for quantification
    base-8, octal for digital computer
    base-2, binary for counting
Binary – represent binary states, e.g., true = 1; false = 0
    true/false;
    exists/does not exist
    value of property: high/low
Special numbers
    π = 3.14...the ratio of the circumference of a circle to its radius
    e = 2.7...
    i = square root of −1
Imaginary number – a real number multiplied by i, e.g., $i26$
Complex – a combination of a real and an imaginary number, e.g., (15, $i26$)
Vector notation combines a pair of real numbers that represents a coordinate pair
Matrix of systems of equations
    $[y_i] = [A_{ij}] * [x_j]$, compressed notation for a set of equations
    e.g., eigenvalues, eigenvectors
Tensor notation describes a linear mapping between algebraic objects
    e.g., stress tensor, strain tensor

TABLE 4.3   The Seven Base Units of SI Form the Basis of Various Combinations of the Metric System of Physical Measures

Symbol	Name	Quantity
s	Second	Time
m	Meter	Length
kg	Kilogram	Mass
A	Ampere	Electric current
K	Kelvin	Temperature
mol	Mole	Amount of substance
cd	Candela	Luminous intensity

Units of Measure

It is imperative that the physical units (Table 4.3) used for measuring and computing on physical quantities are associated with the values of those quantities, but this is hampered by the presence and usage of two competing systems of units of measure: the English foot-pound-second system and the Systeme Internationale (SI) system of base units and derived units. Recently, the International Bureau of Weights and Measures has redefined SI base units in terms of universal physical constants such as the speed of light, Planck's constant, and others, whose values can be determined precisely, anywhere, and at any time. Unfortunately, and despite the advantages of using SI units, the United States has resisted a conversion to SI units, thus burdening American science and technology with the costs of unit conversion. Consequently, computation of physics-based mathematical models requires strict accounting for, and resolution of, differences in the units used to quantify model variables representing physical properties. Some unit ontology resources describe and implement algorithms for translating and scaling measures in one unit into alternative units (Gkoutos et al., 2012). See also the Unified Code for Units of Measure at https://ucum.nlm.nih.gov/.

A great stumbling block faced by biomedical researchers is the diversity of measurement units that range from well-established physical measures such as material amounts (kilograms, pounds) and distance (meters, feet) to domain-specific, indirect measures such as that for blood levels of glycated hemoglobin-A1c (HbA1c). Such measures are defined operationally by, for example, determining the amount of a substrate chemical in a sample compared to the rate at which a known amount of substrate is converted to product under a set of standard conditions. The accuracy and repeatability of such a biochemical "assay" depends on meticulous standardization of reactant concentrations, temperatures, and time measurements. Nonetheless, such bioassays are a mainstay of clinical and research practice that are based on the specificity of particular enzymes for particular substrate reactants. For example, the enzyme glucose oxidase is used in all manner of applications from patient use (e.g., glucose test strips) to clinical and laboratory research.

Property Dimensions

A physical dimension is a generalized quality that represents a measurable aspect of physical reality that is the basis for measurement *units*. Dimensions are a means of expressing the "ground truth" for what the property measures irrespective of the particular unit used to scale the property. For example, the extent of a bone can be expressed in a specific unit of length such as meter or inch both of which have the *dimension* "length". Similarly, degrees centigrade (°C) and Fahrenheit (°F) are units for measure in the *dimension* "temperature". A measure in one dimension is directly comparable only to a measure having the same dimension irrespective of the units of measure.

Re-expression of physical units in terms of their dimensions is a powerful means of type-checking the validity of mathematical expressions

TABLE 4.4   The Seven Base Units of SI Form the Basis of Various Combinations of the Metric System of Physical Measures

Base Dimension		Derived Dimension	
Θ	Absolute temperature		
L	Length	L^2	Area
		L^3	Volume
T	Time	L/T	Velocity
		L/T^2	Acceleration
M	Material mass	M/L	Lineal mass density
		M/L^2	Areal mass density
		M/L^3	Volumnal mass density
N	Moles of substance	N/L	Lineal molar density
		N/L^2	Areal molar density
		N/L^3	Volumnal molar density
I	Electric current	IT	Amount of electric charge

when combining models across multiscale, multidomain physical systems. Table 4.4 shows how (1) seven basic physical quantities with symbols (T, L, etc.) can be combined to form (2) various *derived* quantities that are each a (3) product of base quantities raised to integer exponents each of which takes the form of the general expression, $T^a \, L^b \bullet M^c \bullet N^d \bullet I^e \bullet \Theta^f \bullet J^g$, where a–g are integer values.

Physical quantities can be directly compared only if they have *commensurable* dimensions even if expressed in different units of measure such as meters vs. inches, or pounds vs. dynes. Physical quantities having different dimensions are *incommensurable* and cannot be compared or computed in the same manner. The value in defining a consistent set of base quantities is two-fold. First, such a set provides a formal basis for defining and classifying the myriad of measurement units across multiple biophysical domains, and such basic quantities can be combined as factors and divisors to define the so-called "dimensionless" constants such as Mach numbers to normalize speeds and Reynolds numbers as measures of propensity for fluid flows to become turbulent.

Notations of Scale

Three terms are used to broadly distinguish the broad spatiotemporal scales of entities and processes:

- "Macroscale" describes objects that are a millimeter or larger and is perceptible to the unaided eye, and processes that occur in seconds or longer.

- "Microscale" describes objects that require instrumentation such as a microscope to be visualized, or processes that require, say, an oscilloscope to be displayed in time.

- "Nanoscale" applies to molecular or atomic scale objects whose processes, such as intermolecular reactions or molecular conformational changes, can occur in nanoseconds.

As an example, *macroscale* muscular contraction depends on microscale myocyte contraction, which depends on *nanoscale* myosin binding and unbinding. Modeling multiple scales requires computational integration across scales using factors that are powers of ten, "orders of magnitude" as summarized in Table 4.5, which shows terminology for 21 orders of magnitude.

Some phenomena and their corresponding models are "scale-free" in the sense that observations and models made at one scale apply at other scales. Models of electrical and gravitational fields can apply and are useful across scales from atomic to cellular to astronomical. However, models of fluid flows are very dependent on the temporospatial scales of fluid flow rates and pressures as they depend on the ratio of viscous drag forces versus inertial forces. Intermolecular viscous forces impede the relative motion of atoms and molecules at small scales and can dominate the inertial forces motion of molecules. The viscous force and surface tension that bear on a tiny, say 2 mm, model of a boat on water constrain motions to a few boat lengths. The inertial forces of a full-scale model, however, can dominate fluid viscous drag forces of the water so that a large boat once put in motion will travel many boat lengths before gradually slowing.

TABLE 4.5   Standard Nomenclature for Text and Symbols for Dimensional Scaling Factors Listing Standard Abbreviation (Col 1), Prefix (Col 2), Integer Value (Col 3), and Scientific Notation (Col 4)

G	giga	1,000,000,000	10^9
M	mega	1,000,000	10^6
k	kilo	1,000	10^3
h	hecto	100	10^2
da	deca	10	10^1
(none)	unit	1	10^0
d	deci	0.1	10^{-1}
c	centi	0.01	10^{-2}
m	milli	0.001	10^{-3}
μ	micro	0.000 001	10^{-6}
n	nano	0.000 000 001	10^{-9}
p	pico	0.000 000 000 001	10^{-12}

Normalized and Dimensionless Quantities

Each of the base quantities, above, has a corresponding measurement unit. These are very useful quantities, however, that have no physical dimension in that each is formed as the ratio of two quantities that have the same dimension. For example, "Mach" number is the ratio of the speed of an object through air (e.g., in km/hour) divided by the speed of sound (also in km/hour) under the same atmospheric conditions. Mach number is very useful for aircraft design and testing (but not for subsonic conditions in biophysics) because supersonic airflow is qualitatively different than in the subsonic domain such that an entirely different set of equations are required to model air flows and pressures.

Physical quantities, like mass or the speed of a ball, can be scaled or "normalized" to some characteristic value by dividing each instance of a measure by some other quantity. If that quantity is, say, a standard meter or kilogram then the normalized value is the value now expressed in standard units. In other cases, normalization is important to express how an individual measurement relates quantitatively to measurements of other individuals in a study population. Thus, one can determine how the measure for one individual relates to the same measure for others in the population. For example, the statistical distribution of individual body weights (W_i) in a population can be normalized to a non-dimensional quantity, w_i, by dividing each weight by, say, the population mean weights; $w_i = W_i/W_{mean}$ so that differences in body weight distribution can be compared over time or between populations.

A normalized quantity of relevance to physiology is the Reynolds number, which is the ratio of fluid viscous forces to fluid inertial forces in a flowing liquid. Viscous forces are proportional to the product of a fluid's flow velocity times its viscosity. At low rates, flow is smooth, laminar, and without turbulence. As flow rates increase beyond a Reynold's number value of 1.0, inertial forces exceed viscous drag forces so the flow begins to tumble and become turbulent. This is important for blood flow because turbulence imposes fluid shear forces that inure on vessel linings that accelerates atherosclerosis and triggers clotting that results in stroke and heart attacks.

Extensive vs. Intensive Measures

When comparing or categorizing the various measures used in biophysical modeling, it is sometimes useful to characterize them as extensive or intensive.

Extensive properties such as mass, volume, or total energy of a system are properties that increase or decrease with the spatial extent of a physical entity. For example, the mass of a muscle and the volume of fluid are proportional to the extent of an entire sample.

Intensive properties quantify an attribute at a point within a system but are independent of the spatial extent or the amount of material in the system. Intensive properties such as temperature, fluid pressure, energy density, and material hardness can each be measured and attributed to a spatial point within the bounds of a space-occupying entity. Temperature, fluid pressure, material density, and mechanical stress are examples of intensive properties.

Specific properties are a kind of intensive property that is obtained by dividing an extensive property of a system by the mass of the system. For example, heat capacity is an extensive property of a system. Dividing heat capacity, Cp, by the mass of the system gives the specific heat capacity, cp, which is an intensive property. There are limitations to these categories particularly as entity sizes approach atomic or molecular scales. Temperature and pressure are prime examples. Whereas we can use a probe to measure the temperature at some point in a liter of water, the measured temperature will begin to depend upon the size of the probe if the probe is close to molecular size.

SUMMARY, NEXT STEPS

One aim of biomedical modeling is to bring intellectual and physical rigor to our explanations and predictions of observed biomedical phenomena. Quantitative, physics-based modeling is used to extend and constrain our intuitions and imaginings about how biophysical systems and how well our thought-models actually replicate our empirical observations.

These models have been created over the past 50 years to guide how the multiscale constituents function to control metabolic processes from the cellular to the systemic. Each model has been developed for one, or both, of two purposes. First, deriving and coding a mathematical model of a system and its functionally interconnected parts produces the most explicit representation of a biophysical hypothesis about how the system works; i.e., how each of the parts contributes to the function of the whole. Second, the model provides an analytical platform to "reverse engineer" how systems-level behaviors are explained by the characteristics and connectivity of the system's parts. Next, we turn to the craft of developing and testing computational models of biophysical systems.

CHAPTER 5

System Dynamic Modeling

IN THIS CHAPTER, WE present a few examples of how we describe the foundations of quantitative and qualitative analyses of systems of biophysical processes and their participating continuants. The analytical methods used are those of engineering systems dynamics, a broadly defined, physics-based discipline that aims to account for the operations and behaviors of physical systems. It seeks to use methods of physics and engineering analysis to account quantitatively for flows of material, energy, and information among participating physical entities that constitute a physical system.

System dynamics emerged in the 1950s as a broad approach to representing and analyzing industrial engineering systems and has since been generalized and applied to all manner of other large-scale, highly connected systems in ecology, economics, computer, and electromechanical systems. Mathematical modeling of biological systems is based, often implicitly, on system dynamical principles and practices as developed in engineering and economics. This systems approach provides the foundations of the physics-based biosystem mathematical modeling and analysis. Accordingly, we will describe biophysical systems in terms of stocks and flows of conserved quantities (e.g., atoms, electrical charge), system feedback regulation, signal transduction, and network systems.

System dynamics is an analytical method first developed as a business management tool for representing and analyzing how various business

DOI: 10.1201/9780429469961-5

processes depend on the business stocks, such as raw materials, manufactured products, and the flows of each through manufacturing processes and distribution channels. These basic principles have been generalized for the analysis of physical and engineering systems in terms of stocks and flows of material, electrical charge, and hydraulic flows (Karnopp et al., 2005).

First, we discuss the physiology problems introduced in the prior chapter where physical entities are "things" such as hearts and enzymes that participate in "processes" such as metabolic processing or pumping blood. In the next chapter, we will introduce the perspective that system dynamic analyses and equations and models are expressed using mathematics as the "language" of physics. Mathematical equations are derived for computing the values of key observable system values, such as the flow rate of blood in the aorta, in terms of the values of other measurable, or assumed, physical properties and physical constants (e.g., pi). In Chapter 6, we will describe the Ontology of Physics for Biology using biomedical ontologies and methods for representing biophysical models such as described in Chapter 4.

PRINCIPLES OF SYSTEM DYNAMICS

Both electrical circuits and metabolic pathways can be analyzed using stock-and-flow models in which various kinds of "stuff" are exchanged among "stocks" by discrete "flows" of stuff. In electrical circuits, the *stuff* is an electrical charge which is stored in *stocks* such as a battery or an electrical capacitor. The *flows* are electrical currents by which electrical charge moves from one stock to another while powering electric motors or lights. In metabolic pathways, the *stocks* are stores of chemicals while flows occur as chemical reactions or transport processes in which the chemicals react. Stuff may be water in a plumbing system, cars in a traffic model, electrical charge in an electrical system, or money in a banking system.

Flow systems for chemicals or fluids are modeled as "nodes" representing stocks of stuff and connecting links that represent flows of stuff among nodes. This ubiquitous abstraction is, in fact, so intuitive as to be used without explanation in all manner of physiological system diagrams. However, unlike representational standards for electrical circuits, there are no graphical or schematic standards for biological flow circuits owing to the competing demands of the separate domains such as fluid flow and ion flow. As illustrated in the prior chapter, we are left with a variety of formal standards as well as *ad hoc* approaches ranging from those that are very diagrammatic to those that are quite pictorial and each is a kind of "entity–relationship" ("E–R") or "node–arc" diagrams.

The basic stock/flow modeling paradigm is used throughout the engineering sciences, biomedical sciences, and economics as a way to observe, understand, and predict events in the world. It is based on fairly simple modeling steps:

1. identify the kinds of *stuff* to be modeled – blood, molecules, electrical charge, etc.

2. define *nodes* that are entities consisting of portions or collections of *stuff,*

3. specify *flow paths* by which stuff can be exchanged among *nodes.*

An important aim of this book is to describe how biophysicists generalize basic stock-and-flow modeling for integrated functional analysis of systems across all biophysical domains and structural scales. We will develop and extend a semantic approach that leverages formal ontology methods in Chapter 3 for stock-and-flow modeling based on classical physical laws and computations. Here, the first modeling task is to parse the stuff into nodes and identify the node–node paths by which stuff flows and is exchanged among nodes. For example, consider the following:

Example 1. Vascular blood flow:
 stuff nodes == portions of blood contained in segments of blood vessels
 flow links == blood flow from one vessel segment to the next

Example 2. Enzyme-mediated biochemical reactions:
 stuff nodes == portions of substrate and product molecules
 flow link == catalyzed chemical reactions

Example 3. Membrane electrophysiology:
 stuff nodes == portions of ions in extracellular and intracellular fluids
 flow link == ion flux through the conducting pore of a membrane ion channel

Such biophysical systems are modeled as *flows* of *stuff* among *stocks* of the *stuff.* For example, the stock of blood in the left ventricle flows through the aortic valve into the aorta; the stock of glucose molecules in a cell's cytoplasm flows via an enzymatic reaction, into a stock of glucose-6-phosphate molecules. In each case, the rates of flow (e.g., in mol/sec, or L/min) depend, typically, on the relative amounts of stuff in the source and destination stocks and the nature of the process by which the flow occurs. The flow may be *reversible* such that stuff can flow both ways through the process pathway as for a reversible chemical reaction or blood flow through a simple vessel. Or, flow may be *irreversible* as for an irreversible chemical reaction or blood flow through an aortic valve.

A system dynamic model consists of a set of mathematical dependencies among the existence of the entities and the values of their properties. For example, to be useful and informative, a cardiovascular systems model may not need to represent, for example, the mechanics of cardiomyocyte myofibrils or the transport kinetics of blood cells squeezing through capillaries. To represent all things and processes at all spatial and temporal scales is no model at all because to implement, test, and validate such a complete "model" requires simplifications, assumptions, and representations that constitute models of models. In the following, we will address two challenges of useful biophysical modeling – accurate predictions and hypothesis testing.

Basis in Classical Physics

At a formal analytical level, these are applications of the fundamental theorem of calculus as articulated in geometrical form by the ancient Greeks and subsequently in mathematical notation by Newton and Leibniz in the 18th century (Strogatz, 2019). Subsequent work by a host of theorists, mathematicians, and computer scientists has leveraged these theories into practical analytical methods and applications across the bioscientific spectrum. We adopt limits to both and have excluded from representation and discussion several biophysical domains such as

- Fluid compressibility as flow velocities approach the speed of sound,

- Acoustical and optical physics,

- Psychophysics of aural and visual perception,

- Neural processes of cognition, intelligence, and higher-order brain function.

Even with these exclusions, useful representation and analysis of biodynamic systems requires consistency and consensus agreement not only on the validity and form of the math, but also on the physical meaning of mathematical statements and computations in terms of classical physics. Hence, we offer the OPB as our attempt to provide, use, and test such a resource that spans the domain of system dynamics to represent the physical meaning of the mathematical equations and models as used by computational biologists.

Biophysical modeling has emerged as a craft that is practiced in different disciplinary "silos" of biomedical investigation rather than as a broad discipline with shared views and conventional practices as has emerged in the engineering sciences. This is the central problem we have confronted using available knowledge representation and analytical tools detailed in prior chapters. Although biophysical models are generally driven by curiosity and an urge to understand how biological things and processes work, silo-thinking has left a legacy of incompatible measurement methods, data scaling conventions, mathematical equations, and computational languages.

Conservation Laws

The most pervasive and strongest constraints on system behavior are the so-called conservation laws that are inviolable empirical observations that the amounts of certain quantities within the boundaries of a system cannot change absent the flow of the quantity across the boundary. Conservation laws express fundamental laws of physics that for an isolated system, the total amounts of mass, charge, momentum, and energy are each constant. The amount of matter in a vessel can change only by virtue of the transport of matter across the boundary of the vessel. The amount of momentum of two material entities is conserved during a collision of the entities. The total amount of electrical charge in a device can be changed only by the flow of charge into or out of the device.

Conservation laws are applied to specific cases such as for the portion of a particular kind of molecule contained in the cytoplasm of a cell. Thus, a modeler will derive the so-called "mass balance" equations to express that the rate of changes of the amount of matter (as glucose) in a cell equals the sum of (1) the net rates of glucose synthesis and consumption, and (2) the net rate of glucose uptake and export across the cell's bounding membrane. Similar conservation laws for electrical charge, momentum, and thermodynamic energy are routinely represented as constraints within models.

Stocks and Flows

A *stock* is a portion of *stuff* – such as a portion of blood, protein, or electrical charge – where the portion has an observable *amount property* that is a measure of the amount of stuff that is a part of a system at a moment in time. Amounts of stuff are measurable *state variables* such as volume, mass, weight, or chemical molarity. The *flow* of stuff is the transfer or transport of stuff at a particular *temporal rate* (e.g., in mL/min, lb/sec) that is defined in terms of the temporal rate of change or flow rate (e.g., moles/sec) of stuff from one stock to another through a pipe or a membrane ion channel. Thus, the amount of stuff in the source portion is diminished and that in the destination portion is augmented. This is the simple essence of stock-and-flow modeling, which engineers and biophysicists have generalized to represent and compute on all manner of real and theoretical systems to express models and make predictions. In the following, we will describe examples and generalizations that have been used for biosystems analysis.

System State

A *state variable* is one of a set of variables that characterize the physical state of a system at a moment in time. Such counts may be an *integer* count of individual dollars or animals (e.g., 1,234 dollars or 350 animals) or, in most biophysical systems to be considered here, amounts must be expressed in *real* numbers that may, or may not, be *integers*. For example, whereas the 23 chromosomes are readily counted, the number of glucose molecules in cytoplasm can only be expressed as continuous real numbers in units of concentration as moles per liter, such as 100 mM/L that may be convertible to other units such as milligrams per deciliter (mg/dL) or per cell (mg/cell).

Partial Differential Equations (PDEs) or Finite Element (FE) Analysis

One might suppose that the most exact analysis of continuum entities such as bone bending or blood flow would be to compute the motions of each atom that participates in the process. However, the computational costs would be prohibitive for even small entities of biomedical interest. For models and materials at larger scales, a more usual approach is to ignore the atomic/molecular structures by assuming that material, forces, etc. are each continuously distributed in space – the "continuum" assumption. Thus, one can assume that every physical property (material density, mechanical stress, temperature, etc.) has a particular value that varies continuously in space. Given this representation, it is possible to derive a set of PDEs that can be solved exactly but only for simple structures and geometries. For the most part, investigators turn to the so-called finite-element (FE) models.

Discrete, stock-and-flow models are vast simplifications of what actually occurs in the three-dimensional flow field of a real biological system. For example, a single, discrete variable, Fa, may be defined as the overall total flow rate of what is actually a spatial distribution of flow rate densities that apply to each point in the viscous fluid through a conduit such as orifice of the aortic valve.

Ordinary Differential Equations (ODEs)

The essence of system dynamic analysis is to mathematically solve or computationally simulate how flows of "stuff" change the amounts of stuff in various stocks. For example, the amount of water in a reservoir (i.e., a *stock*) changes as the net rate of *flow* into and out of the stock. Assuming water to be incompressible, then the change of amount during a time interval, Δt, will equal the net difference between rates of inflow (Fin) and outflow (Fout) times the duration of a time interval, Δt, which can be represented algebraically as

$$\Delta V = (\text{Fin} - \text{Fout})\, \Delta t.$$

Such relations can be derived, evaluated, and tested as part of, say, an experiment that measures volumes as discrete values (e.g., 10.2 mL, 22.6 mL, 31.7 mL) at discrete time interval, Δt (e.g., 12 sec, 16 sec, 25 sec) as may be tabulated on a data sheet. We can recast the above *finite difference equations* as a proper *differential equation*:

$$dV/dt = \text{Fin}(t) - \text{Fout}(t).$$

Such a differential equation is exact and true by the definitions of the rate of volume change at an instant in time is dV/dt expressed as the difference between inflow and outflow rates (Fin(t), Fout(t)).

Systems Analysis Illustrated

Engineers have long understood the critical need to simulate and analyze their designs to assure the successful design and deployment of their mechanical, chemical, and electronic contrivances. Engineering design, testing, and development depend on tools for analysis and synthesis as two approaches that are broadly combined as "systems analysis" a tool for representing and modeling networks of interacting discrete entities such as plumbing systems composed of pipes, mechanical system of ropes and pulleys, or electronic circuits consisting of wires, resistors, and capacitors.

In each case, system analysis maps the flow of some conserved substance (e.g., matter, charge, or energy) that is distributed among nodes that are interconnected by process flow pathways. For example, the top panel of Figure 5.1 illustrates how the flow rate (F) of blood from the proximal aorta into the distal aorta is proportional to the pressure differential $(P_1 - P_2)$ from proximal to distal aorta ($P_1\_P_2$; top panel, Figure 5.1) that exists along some conduit (equation to the right). However, this simple, two-compartment analysis may not have sufficient spatial resolution to accurately account for the flow and pressure differentials as measured in a dataset. For example, one may need to account for pressure/flow changes due to vessel tapering or plaque blockages, in which case one may need to partition the vessel into many segments each of which has its own blood volume and flow connections with the previous and following vessel segments (middle panel, Figure 5.1). Such a flow model could account for the effects of local obstructions or vessel taper on blood flow and pressure profiles using *ODEs*. If one is availed of continuous measures of cross-sectional area along the vessel, then one can push the analysis further based on PDEs.

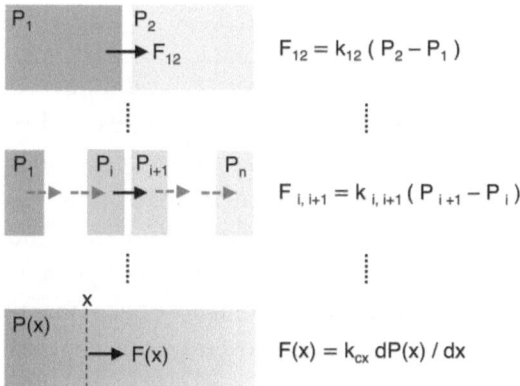

FIGURE 5.1 Three levels of analysis of the diffusion of some substance along a spatial pressure gradient pathway. (Top) A simple two-compartment model, (middle) partitioning of the pathway into n discrete pathway "finite" elements, and (bottom) the eventual conversion of the discrete partitions into a continuum model that can be analyzed using differential equations.

Continuum entities may not have natural demarcations as can be used partitioning into discrete components, or they may be composed of so many elements as to be considered to be already spatially continuous. Such entities may be analytically partitioned into multiple "FEs" each of which is a small part of the whole. For example, an analysis of how an elastic cable (e.g., a tendon) stretches might be segmented into numerous FEs that are each, say, 1/100th of the cable total length.

A FE analysis may begin with a representation of an entire entity, such as a piece of bone or a volume of blood, into many – tens to thousands, typically – spatially distinct elements that interact physically according to, for example, the structural mechanics of stresses and strains, or fluid dynamics of pressures and flows. FE analysis may be required for solving certain problems but it is overkill for most as it can require extensive computational resources, substantial costs, and data that is difficult and costly to acquire. However, FE analyses do contribute to our understanding of many physiological phenomena, particularly heart contractile function, vascular blood flow, and bone mechanics.

One of the key tenets of system dynamical analysis is the reduction in complexity that stems from decomposing complex systems into physical and functional subsystems to support analysis of biophysical systems, their behaviors, and their functions. In the following, we apply the system dynamics as a representational framework for analyzing cell biophysical entities and processes.

SYSTEM PROCESSES

The biomedical information (Chapter 2) and knowledge (Chapter 3) resources define, represent, and organize the bits and pieces of biological structures (continuants) and processes (occurrents). These valuable resources are available yet they only incidentally represent or enforce the biophysical principles by which such systems function. Thus, neither memory models nor mimicry models specify the biophysical mechanisms that are components in a biophysical system whose behaviors they are intended to imitate. Here, we will describe several such component biophysical mechanisms and describe functional features by analogies to more familiar electrical, fluidical, and mechanical mechanisms. Recognizing and formalizing such analogies, as will be extensively discussed below, is the basis for recognizing and formalizing a semantics of physical systems that are the basis for Ontology of Physics for Biology as described in the next chapter.

Metabolic Flow Processes

Biological physical entities are composed of atoms and ions that are bound together to form molecules that can aggregate to form cells and their organelles. Such objects can be studied as an individual object, as countable sets of objects, or as an uncountably large set of objects. For example, the bending and twisting of a single (representative) protein molecule may be modeled in a molecular dynamic simulation to identify the various conformational states (shapes) that can occur and how likely are the transitions between the various shapes. Alternatively, depending on available data and analytical needs, one might model an aggregate of such proteins and be concerned with how many individual proteins are in one structural state versus some other state.

Chemical Reactions

Chemical reactions are a critical and heterogenous set of cellular processes that are central to cellular metabolism and the regulation of virtually all biophysical processes. Some of the reaction processes are a simple isomerization whereby a molecule switches between two stable shapes or conformations, or a dimerization whereby two molecules stably bind together to form a molecule of a different type. A simple version of such reactions obeys the law of mass action whereby, to a first approximation, the chemical reaction rate is proportional to the product of the chemical concentrations (or, more precisely, the chemical activities) of the reactants that are modeled by the so-called "mass-action" reaction rate laws of the form, $R_{AB}=K_{AB}$ [A] [B]. Such a rate law applies to reactions that are at, or near, chemical equilibrium; i.e., when forward and backward reaction rates are approximately equal.

However, the equilibrium assumption does not hold for most cellular metabolic and transport processes and, in particular, for mitochondrial and electrophysiological applications that require more complex models based on non-equilibrium, chemical thermodynamics (see, for example, Beard and Qian, 2008). Here, we will, however, refer only to examples of near-equilibrium chemical reactions and mass-action kinetics.

Enzymatic Reactions

Chemical reaction flow modeling is applied to simple and to large and very complex and large biochemical reaction systems spanning simple conformational state changes to those that include pathways involving multiple enzymes and reactants as well as key metabolic control and branch points.

A cellular metabolic pathway is a series of discrete biochemical reactions such as a reaction that converts a "substrate" molecule (S) to a "product" molecule (P) that are coupled and can be analyzed using basic chemical reaction kinetics tells us that the reaction rate ("velocity") is proportional to the concentration of S; i.e., $V_{SP} = dP/dt = kS$, where S and P are the concentrations of S and P, respectively. Although such "mass-action kinetics" may be suitable for maximizing the throughput of industrial processes, cell biological reactions require catalysts to accelerate and regulate complex biochemical reactions that otherwise occur only rarely and slowly. Indeed, at the turn of the last century, it was understood that protein enzymes served as catalysts (Figure 5.2) to accelerate the S-to-P reaction rate, V_{SP}, is proportional to both S and P, but as more S is added the S → P reaction rate reached a maximal rate, i.e., "saturated", that is denoted as V_{max}.

The starting point for analyzing complex biochemical systems within cells began with Leonor Michaelis and Maud Menten, who, in their seminal work from the early 1900s, explained that the enzyme, although the key accelerator of reaction rate, also participates in the reaction's rate-limiting step, which is evidenced as an apparent saturation (limited) of the reaction rate. They derived a simple mathematical model of enzyme-mediated reactions that solved a biochemical problem. In basic chemistry, we learn that a simple reaction whereby molecules of A are converted irreversibly to molecules of B, the conversion rate is proportional to the concentration of A. This is distinct from simple mass-action kinetics whereby the rate or velocity V_{AB} of the A-to-B conversion reaction is proportional to the concentration of A times a reaction rate constant, K_A; i.e., $V_{AB} = K A$.

Consider the classical Michaelis–Menten (MM) model of enzyme kinetics as derived in their classic 1913 paper (translated in Michaelis et al., 2011). The MM model (Figure 5.2) consists of a set of linked chemical reactions whereby a molecule of a substrate (S) binds to a molecule of a

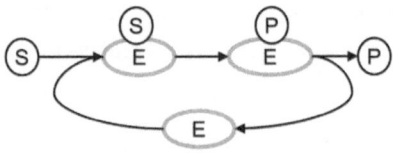

FIGURE 5.2   Michaelis–Menten molecular state model wherein an enzyme molecule (E) binds a substrate molecule (S), catalyzes its chemical conversion to a product (P), releases P, and then restores itself to its original, unbound state, E.

catalytic enzyme (E) to form a "substrate–enzyme complex" (S–E'). Once bound to S, the enzyme catalyzes the conversion of S to a product, P, which dissociates from E and recycles to catalyze the remaining substrate. What distinguishes this model from a mimic is that each variable in the kinetic equations represents a measurable physical property of a physical participant: chemical *concentrations* of the molecular participants and *rate constants* for the chemical reaction among the molecules.

The state diagram and equation, above, explain key kinetic features of this simple and nearly ubiquitous reaction schema. For example, a plot of reaction velocity, Vsp, as a function of S as plotted below shows that V is proportional to S at small values of S but V saturates at the limit, Vmax, at large values of S that approaches and surpasses Km according to the classical Michaelis–Menten rate equation and its graph (Figure 5.3).

$$Vsp = Vmax * S/(S + Km)$$

The saturation of reaction velocity is a key characteristic of all enzyme-catalyzed reactions because it takes a finite time for binding, catalysis, and unbinding so that fewer enzyme molecules are free to initiate reactions and contribute to the reaction velocity. This functional theme has a myriad manifestation including hundreds of fully analyzed reaction kinetic schemes analyzed by Segal in his compendium of hundreds of variations on enzyme kinetic models including multiple catalytic sites, sequential catalysis, "hybrid Theorell-Chance ping-pong", and "iso ping-pong" systems (Segel, 1975).

Models of the saturation kinetics of enzyme catalysis provide an important model-as-metaphor for all manner of physiological processes,

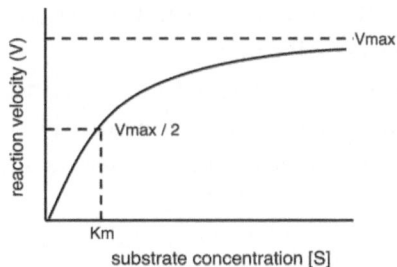

FIGURE 5.3 Plot of the Michaelis–Menten equation showing how reaction velocity, V, increases with increasing substrate concentration, S, until the enzyme rate reaches a maximum, Vmax.

which exhibit saturation of process rates due to limits on process participants. Oxygen uptake via the respiratory system, work output by muscles, and nerve firing rate all show saturation kinetics, analogous to enzyme kinetic saturation, due to mechanisms of competition for limited resources.

Fluid Flow Processes

In Chapter 1, we described key characteristics of blood flow through the circulatory system that include large differences in spatial scale of the blood flow path from the heart ventricles, through the aortic valve, and through progressively smaller arterial vessels into the capillaries. We noted a progressive lessening of both pressure gradients and flow rates. These present major challenges to physiologists and clinicians in understanding how blood pressure is regulated. Here, we illustrate some of the modeling challenges of this domain.

A kindergarten physicist would recognize the left ventricle as a balloon containing a fluid (blood) whose pressure increases as it contains more blood, and the aortic valve an outflow path analogous to a drinking straw that resists flow as its diameter decreases. Key dynamic properties of such a system can be demonstrated with a simple equation with linear assumptions. A simple experiment to perform and model is to inflate the left ventricular "balloon" with blood, open the aortic valve "straw", and measure two things: aortic valve blood flow rate (e.g., in mL/sec) and how quickly ventricular blood pressure (e.g., in mmHg) drops as functions of time. We begin with a simple model assuming linear dynamics and then discuss representing various nonlinearities that are so characteristic of biophysical systems. For example, the flow rate of a "Newtonian" fluid through an orifice is inversely proportional to the cross-sectional area of the orifice and to the viscosity of the fluid; i.e., the flow rate is lower for a more viscous fluid flowing through a small orifice.

Whereas such assumptions are routine in engineering systems, bioengineers must contend with a number of factors. First, blood is a distinctly non-Newtonian fluid whose flow viscosity depends on the flow rate itself. Second, aortic valve diameter and, thus, blood-flow resistance, can change as blood expands the valve orifice. Ventricular pressure does not vary linearly with the volume of blood it contains due to the nonlinear elastance of the ventricular myocardium. However, refinements such as these can be incorporated, as needs be, in a progression of more sophisticated, more accurate models.

Steady Blood Flow – Ohm's Law

Ohm's law was originally derived for electrical systems to represent the relationship between an electrical current that flows through an electrical-conducting device, and the electrical potential difference across the device. Ohm's law expresses the relationship of values of the gradient of electrical potential (E) and electrical charge flow rate (I) for current flow through a conducting pathway (e.g., a wire). For most engineering applications, Ohm's law is a linear relationship between values of I and E according to the pathway resistance (E=IR) or its conductance (E=I/G). To be developed below, Ohm's law and its elaborations find important applications in the electrophysiology of cell membranes but let us first examine its applications to modeling air and blood flow.

Pulsatile Blood Flow – "Windkessel" Model

"Windkessel", in German, means "air chamber" such as used to damp out pressure pulses of a mechanical, piston fluid pump (Westerhof et al., 2009). Imagine inflating a leaky air mattress so that each puff of air inflates the mattress but only transiently as escaping air deflates the mattress. The same applies to pulsatile blood flow by which blood is pumped through a series of blood vessel segments each of which behaves like a "windkessel" that is periodically inflated during systole by upstream blood flow pulses and are deflated during diastole as the blood flows into the down-stream segment. As a consequence, pulses are both delayed and diminished as they propagate through a vascular tree until, at the level of the terminal arterioles, there is very little detectable pulse.

Thus, the entire arterial tree can be decomposed into a branchwork of windkessel models. Based on this idea, vascular physiologists have, for years, used variations and elaborations of the electrical circuit elements shown below as a computational analog of a fluid dynamical windkessel for a vascular segment (Figure 5.4).

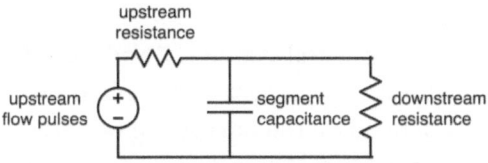

FIGURE 5.4   Schematic "windkessel" model of pulse-to-pulse arterial blood flow through a single vascular segment in which input pulses are damped and diminished as blood is pumped through an arterial tree.

Atherosclerotic vessel segments can be modeled by increasing the flow resistances through zones of atherosclerotic plaques that occlude flow and decrease vessel compliance. Elaborations of the windkessel schema can include "inductor" elements to represent the inertia of blood flowing in the larger vessels. The basic windkessel model can be elaborated for particular modeling needs. For example, flow resistances (i.e., the R's) for blood are substantially nonlinear due to the interactions and mechanics of the various blood cells and macromolecules. Also, inertial effects seen due to systolic pressure pulses diminish as blood flow rate and pulsatility diminish in the more distal vessels. The value to modelers of the windkessels modeling is that a broad range of normal and pathological variations can be represented by simple parametric variations of simple mathematical models.

Hagen–Poiseuille Flow Law

Ohm's law for fluid flow can be applied to flow paths in systems where the dependence of flow rate on pressure gradient can be empirically characterized by a single parameter, R. However, Ohm's law can be elaborated for fluid flow as the Hagen–Poiseuille law that is a *constitutive* relationship between the physical properties of the fluid and the geometry of its flow path.

$$Q = \Delta P \ A^2 / \pi \ L \ \mu$$

where

Q = volumetric flow rate through vessel segment,

ΔP = pressure drop across segment,

A = vessel cross-sectional area,

L = vessel segment length,

μ = coefficient of fluid viscosity.

By explicitly representing vessel geometry and fluid viscosity, this constitutive law can be used to represent and explain how, for example, the blood flow rate may depend on the narrowing of a vessel due to the buildup of atherosclerotic plaque. It is significant, therefore, that the flow rate, Q, is very sensitive to such narrowing as it is proportional to the square of area, A, so that a flow rate, Q, is reduced to 25% by a 50% narrowing of the vessel.

The Hagen–Poiseuille law is expressed in terms of a coefficient of viscosity (μ) that applies to the so-called "Newtonian" fluids (e.g., water, air) whose viscosity is independent of flow velocity. However, the viscosity of biological fluids can be very dependent on flow velocity, vessel size, and fluid composition such as hematocrit. For example, blood flow viscosity increases with increasing blood hematocrit, protein concentration, and coagulation state. In the extreme, for example, blood cells must queue, one-by-one to squeeze through capillaries that must dilate to accommodate cell movement.

Ion Flow Processes

All cells maintain a "membrane potential" – that is measureable as the electrical voltage across their cell membrane, Em – as measured relative to the surrounding extracellular space, may range from a few millivolts to −100 mV whereby the voltage is negative relative to the extracellular space. Although only about 1/15th of the voltage of a typical flashlight battery, the 100 mV exists across a cell membrane that is, however, only about 0.1 μM thick so that the actual voltage gradient is on the order of 100,000 V/cm that is large enough to dramatically control molecular components of the cell membrane. Thus, the field of electrophysiology is dedicated to studying how cells and tissues and organs control and exploit this voltage for or information transmission by nerves, hormone secretion by endocrine cells, and contraction by muscle cells.

Origin of the Membrane Potential

Here, we present a very brief overview (Figure 5.5) of the ionic basis of the membrane potential, its control, and its role in triggering membrane electrical activity that is a key controller of muscle contraction, neural signaling, and hormone secretion.

Similar active ion pumps and passive ion transport exist in virtually all cell types but key cell types deploy specific combinations of ion channels to subserve specific cell functions.

Charging the Membrane's "Ionic Batteries", E_K and E_{Na}

The membrane Na-K ATPase (NKP, circle) "ion pump" establishes transmembrane gradients (wedge shapes) for Na$^+$ and K$^+$ using metabolic energy from the hydrolysis of ATP-to-ADP. This process is analogous to charging electrical batteries by injecting electrical currents to transport ions (e.g., Li$^+$) across the dielectric charge separator in the battery. The resulting

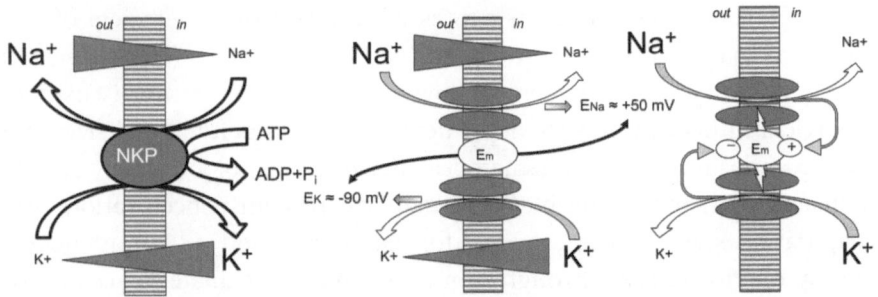

FIGURE 5.5   Origin and modulation of the membrane electrical activity depends on three basic processes: (left) metabolic energy of ATP is consumed to "pump up" transmembrane Na+ and K+ concentration gradients to charge the "ionic battery", which (middle) polarizes the cell's membrane potential, Em, (right) which then depends on relative rates of Na influx and K efflux that determine the membrane potential that controls channel activity.

membrane potential is negative relative to the surrounding extracellular space that is taken, by convention, as the electrical ground (i.e., zero millivolts, mV).

Setting the "Resting" Membrane Potential

At any time, the actual membrane potential lies (see diagram, above) somewhere between the "equilibrium potential" (Vm or Em) is defined as that membrane potential as an electrical force just balances by the diffusion gradient for the ions according to the "Nernst potential" for Na K ions can be computed as at any point in time by the balance of regulated between the equilibrium potentials E_{Na} (usually, $\approx +50\,mV$) and E_K (usually, $\approx -90\,mV$) according to the relative permeabilities to Na+ and K+ ions. These very basic electrophysiological mechanisms are expressed and elaborated in all manner of cells and are critical processes in many functional systems.

All-or-None Action Potentials

A key regulatory phenomenon in virtually all cells is the establishment and control of the electrical potential (voltage) across the cell's plasma membrane. It is a key regulatory property by which a cell responds to its local, extracellular metabolic, hormonal, and neural inputs. Typically, a cell's membrane potential may vary between −50 and −80 mV relative to the potential of the extracellular fluid which, by convention, is taken as

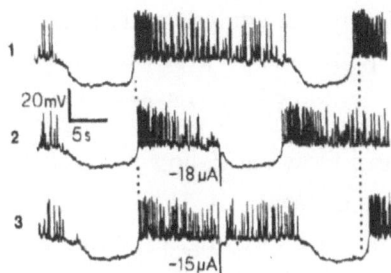

FIGURE 5.6   Electrical suppression of membrane electrical activity that mediates calcium ions into insulin-secreting pancreatic beta cells.

the zero, ground potential. Reduction, or depolarization, of the membrane potential is a key regulatory event that can trigger the opening of calcium-ion channels in the plasma membrane leading to the influx of calcium ions and the activation of cellular secretion and contraction.

Patterns of Spiking and Cell Activation

Virtually, all cells have ion channels and many are specialized to organize their ion channel currents to mediate a range of membrane electrical activity that participate in processes of cell–cell signaling, secretion, and contraction. In such cases, the analytical task is to model the: rates of calcium uptake into the cell cytoplasm, of the activation of the particular cell process, and of the transport of calcium into intracellular stores and back into the extracellular space. The challenge is, for any particular cell, to understand and quantitatively describe the relevant electrophysiological and membrane transport processes. For example, mouse pancreatic beta cells have a "bursting" pattern of membrane electrical activity (Figure 5.6) that mediates the uptake of calcium ions that triggers insulin secretion.

SYSTEM DYNAMICS MODELING

We have discussed an array of computational approaches representing and analyzing a variety of biophysical systems, entities, and processes. In each case, the content, execution, and predictions of the models must be based on a common understanding of the physics underlying the physical phenomena. Here, we will outline such a view of classical physics that supports predictive computational dynamical models. First, we will describe the temporal and spatial scales of the contents of informatics resources that are relevant to biophysical analysis and modeling. Next, we define and discuss the physical quantities by which biophysicists observe, measure, and quantitative aspects of the biophysical entities and processes.

Principles of Stock-and-Flow Modeling
States of Stocks – Rates of Flow

At any moment in time, various things exist in a system, such as the blood in a vessel, the glucose in a cell, or the muscle in an arm. The physical state of such a system is defined and observable in terms of measures of *amount* properties such as the *volume* of blood, *moles* of glucose, and the *tension* of a muscle. As biophysical processes *occur*, these state properties change; understanding a system requires an understanding of how such changes occur in terms of causes and effects, and accounting for the temporal rate at which they change.

Modeling States and Rates

We now introduce a minimal vocabulary for stock–flow modeling as required for representing experimental observations and explanatory hypotheses. In the remainder of this book, we will develop and extend this approach as a general method for representing complex, multiscale, multidomain systems. We start here to observe that basic stocks and flows are organized into networks whose topological organization becomes the basis for explaining and testing complex systems behavior.

The conservation relations are universal and provide critical constraints on quantitative modeling and are foundational to fields as disparate as biophysics, particle physics, and astrophysics. In biological modeling, they are the basis for continuum models of fluid flow as in blood circulation, biomolecule transport, nutrient metabolism, ion fluxes through cell membranes, and gas transport in the lungs.

Furthermore, they apply to the bookkeeping of discrete things such as sets of human beings in a room, cells in a bit of tissue, or protein molecules in a cell's membrane. However, biological things such as cells and biomolecules can be created and/or destroyed because processes may merely rearrange, otherwise conserved, atoms, charges, and energy. Thus, a protein of a particular sort is synthesized by linking existing amino acids into chains. Two cells may bifurcate into two cells by redistributing the material and energy in the parent cell without violating conservation laws of mass and energy. In these cases, the number or count of a set of particular objects, cells, or molecules, can change without violating the fundamental conservation laws.

Here, we discuss examples of the classical laws of physics, each of which dates from the late 19th century but appears and reappears in models in all biophysical domains: Ohm's law, Faraday's law, and Newton's first law. We will first present the three classical laws as presented in college physics and then restate the laws in a more general form that anticipates the formal semantics of biophysical systems as developed in the next chapter. We will discuss specific rate laws below.

Classical Laws

The conservation principles and their mathematical expressions (e.g., $V = IR$ for Ohm's Law) provide important constraints on how the amount of stuff in a node depends on the biophysical flow processes that link nodes. While overarching conservation principles apply to each kind of stuff in a system, the "rules" that govern node–node flow rates differ substantially in complexity across biophysical domains and may include superimposed regulatory mechanisms. How these features are represented and computed in a particular model depends on the scope of data to be modeled and the questions to be answered.

Conservation and Continuity – Constraints on Amounts

One of the foundational principles of physics is the so-called "conservation laws" that express the unviolated empirical observation that the amounts of certain quantities – matter, charge, energy, momentum – that exist within an enclosed system can be neither created nor destroyed. For example, the amount of matter – number of atoms, gallons of water, calories of energy – must be conserved within the boundaries of a closed system that neither imports nor exports the quantities across the boundary of the system. Thus, any change in the amount of conserved stuff must be due to a flow of stuff into (f_{in}) or out of (f_{out}) according to the so-called "continuity" relation:

$$\Delta S = (f_{in} - f_{out}) \, \Delta t$$

In addition to the conservation of matter, parallel conservation principles hold for the *amounts* of other conserved quantities – the *amount* of electrical charge, the *amount* of mechanical momentum, and the *amount* of thermodynamic energy. Thus, for each of the four conserved quantities, there is a conservation equation of the form above that applies and can be used and must be satisfied for each of the conserved quantities.

Dynamical models depend explicitly on such conservation equations to assure completeness and consistency of equations for a flow system by periodic error checks to assure that the sum of conserved amounts does not change during computations. Such bookkeeping is especially difficult in complex metabolic systems in which atoms are always changing roles as molecular partners but can never leave the dance. For example, one might add up the number of carbon atoms in the molecules participating in a metabolic sequence to assure that the stoichiometry of reaction sequences is properly accounted for. Comparable accounting of thermodynamic energy could be implemented for the reaction sequences.

Ohm's Law – Flows and Forces

We have introduced electrical applications of Ohm's law as a linear proportionality between electrical current (I, a flow rate) and the voltage difference driving force (ΔE, a force differential) for an electrical circuit pathway. For example, first, the volume flow rate (f) of blood through the aortic valve can be modeled as proportional to the pressure difference of blood pressure at the inflow from the left ventricle (p_V) minus the pressure in the at the outflow (p_A) into the aorta, divided by a flow resistance parameter (Rv) for flow through the valve (as below).

$$f = (p_V - p_A)/Rv$$

Second, for our simple model let us assume that the ventricular blood pressure (p_V) is proportional to the volume of blood in the ventricle (V) times the elastic stiffness (k) of the ventricle due to tension in its wall. That is, the more blood in the ventricle, the higher the ventricular blood pressure, p_V.

$$p_V = kV$$

For simplicity's sake, let us now assume that there is no backpressure from blood in aorta so that

$$p_A = 0$$

Under these conditions, we expect blood to flow from the ventricle to the aorta at a rate that is initially high but approaches zero as the ventricular blood is exhausted and ventricular pressure goes to zero.

Third, we express a *differential equation* that equates the rate of change of ventricular volume (dV/dt) to the aortic valve blood outflow rate (f) according to

$$dV/dt = -f$$

If outflow is driven by ventricular pressure but is impeded by the valve's fluid flow resistance, then

$$dV/dt = -p_V/R$$

If p_V is related to the elastic, balloon-like expansion of the ventricle, then we can recast the *differential equation* in terms of a single dynamical variable, V, and two constants, k and R:

$$dV/dt = -V \ (k/R)$$

Faraday's Law – Amounts and Forces

Faraday's law was originally derived for electrical systems to represent the relationship between the amount of electrical charge (in coulombs) stored within a device and the electrical potential (in volts) that is required to maintain the charge difference across the device. In the domain of electricity, a "capacitor" device consists of two electrically-conducting plates or sheets separated by a non-conducting "dielectric" plate or sheet. In the domain of electrophysiology, the charges are comprised of ions (e.g., Na+, K+) dissolved in fluids separated dielectric membranes that bound cells and intracellular membranes.

In the domain of fluid dynamics, such as for aortic valve blood flow, we can view the left ventricle as a fluid capacitor in which the ventricular blood pressure (a force) is proportional to the amount of blood in the ventricle according to the elasticity of the ventricular wall.

Newton's Law – Forces, Accelerations, and Momenta

According to Newton's law, a material entity of mass (m) is accelerated (a) in proportion to the net force (f) that is applied to it. This law can thus be expressed as f = ma, or by noting that acceleration is the rate of change of velocity, i.e., a = dv/dt, we get the equivalent form of Newton's law as f = mdv/dt. In the fluid domain, as for our blood flow example, Newton's law is used to calculate the acceleration of blood flow rate through the aortic valve according to the pressurization of blood due to muscle contraction in the ventricular wall. Similarly, Newton's law is used to compute the accelerations of bones and tissue due to contractile forces of skeletal muscle.

These basic physical laws become our jumping off point for analyzing complex biological systems whose representation, functional interpretation, and cause–effect analyses are dominated not only by the complexity of systems but also by the non-linearities of the basic engineering and physical principles.

Systems Pathway Perspective

The topological features of a functional network may begin to explain patterns of a system's behavior and suggest its function. Cause–effect relations can be inferred or discounted according to recognized qualitative pathway patterns that may account for experimental observations, but deeper understanding requires quantitative methods based on observations and computations about the measurable, quantitative attributes of the stocks and flows that constitute a system. Whereas these graphical and topological elements are routinely used to describe problems in the biophysics and biomedical domains, these "models" are only conceptual, qualitative descriptions that have only an implicit representation of the quantitative physical properties that are the observable *attributes* of experimental systems.

Furthermore, there is a particular emphasis on identifying and defining motifs that are "pathways" that consist of multiple motifs that are functionally linked in a signaling or metabolic system. For example, the functional significance of patterns of pathways and motifs depends critically on the quantitative values of the physical properties and parameters that describe motif elements. Such biochemical and cardiovascular diagrams are very valuable topological maps of systems – biochemical reaction fluxes or cardiovascular blood flows. They can be used in graphical form to guide thinking in their respective domains but can only faintly represent multidomain interactions such as biochemical impacts on blood flow (or vice versa) or interactions with, say, the activity of neural systems or the musculoskeletal system.

Glucose Homeostasis

One of the key features of physiological systems is their capacity to self-regulate around some healthy operating point where various physiological metrics such as blood pressure and metabolite levels hover around normal levels. For example, there is a long history of modeling processes of insulin secretion, cellular metabolism, and blood sugar regulation as control systems and for using artificial devices as therapeutic substitutes.

Blood Pressure Control

As emphasized in Chapter 1, chronic, essential hypertension is a failure of normal blood pressure control whose physiological origin is only poorly understood but involves multiple, anatomical entities and multiple interacting physiological control pathways (Figure 4.2). For example, blood pressure is sensed by "baroreceptor" nerves in the walls of various blood vessels. Baroreceptor signals are sent to brainstem regulatory centers whose neural signal controls the frequency and strength of heartbeats and, thus, cardiac output flow rate, and the flow resistance of arterial vessels.

As an example of the application of system dynamics, Canetea and colleagues present a comprehensive, high-level map of systemic control of blood pressure that includes neural and hormonal pathways by which blood pressure reading is transmitted to brainstem regulatory centers (Canetea et al., 2019). The model represents blood pressure control in terms of five key physiological property measures: aortic pressure, heart rate, ventricular elastance, (vena) cava elastance, and systemic (flow) resistance. Each such value is compared to a "reference" value and the difference ("error") value is fed to a "transfer function" (H_i) that determines the values of the observable measurable factors such as heart rate and ventricular elastance.

Homeostasis as an Emergent Property

Homeostasis – the steady maintenance of physiological function – is a quintessential feature of biological life and function. Organisms can only survive and reproduce if their structures, processes, and properties are maintained within the operating limits established by evolution. Claude Bernard coined the term "internal milieu" to describe such stability at cell and biochemicals and recognized the operation of nested feedback mechanisms.

Clinical "vital signs"– body temperature, heart rate, blood pressure, blood oxygen level, and so on – and "panels" of clinical laboratory values – blood sugar, hematocrit, electrolytes, etc. – are "vital" because they are maintained at very stable values within narrow limits because they affect so many bodily functions. Deviations from baseline reveal or contribute to ongoing pathological processes and are cause for concern as in fever, hypertension, hypo- and hyper-glycemia, and others. To explain and understand such examples of homeostasis, physicians, physiologists, and modelers often invoke in common parlance or in model code, the language of engineering control systems. However, there is a profound difference as there is no discernible "setpoint" value in physiological control systems.

A temperature or speed target value is set as a physical analog – a voltage or mechanical position – to which a corresponding analog of the actual, measured value is compared. The difference between target and actual values becomes an *error* signal that changes an effector such as a fuel flow. Thus, engineered controllers are designed to minimize the error as the difference between an *actual* measured value (e.g., 65° or 75 mph) and a *set point* value (e.g., 72° or 80 mph). Models of such setpoint controllers may be best classified as "mimicry" models because they function "as if" there is a measurable setpoint value in the physiological system.

Such set-point behavior is the so-called "emergent" property of complex dynamic systems as typified by physiological regulatory systems. For example, there is no reference level of glucose, or any other biochemical species, within a beta cell to which the prevailing extracellular or intracellular glucose is compared. Rather, the "setpoint" is a virtual property defined only by the system's behavior. The "setpoint" is, however, a useful empirical measure in diabetes, obesity, and hypertension because in each case the system is behaving as if it is a setpoint controller but, in each case, the setpoint has been changed so that regulation of blood sugar, body mass, or blood pressure is still occurring but does so around apparent setpoints that are elevated.

Thermodynamic Constraints and Energy Dissipation

Physiological systems are physical systems and, so, are subject to the same rigorous thermodynamic constraints on energy use and on entropic energy dissipation as govern engineering and other physical systems. Thermodynamics is an explicit consideration in the study of the physical chemistry of enzyme catalysis (Goldberg et al., 2006), cell system processes (Beard & Qian, 2008), metabolic systems (Beard et al., 2002; Qian & Beard, 2005), and contractile systems (Beard & Qian, 2008).

More globally, thermodynamics is the basis for analyzing large-scale, multidomain physiological systems using network thermodynamics (Katchalsky & Kedemo, 1962; Oster et al., 1973) and, more recently, bond graph analysis as applied to multidomain physiology (Gawthrop & Crampin, 2018; Le Rolle et al., 2005). At broader scales, entropic changes due to energy dissipation are understood to be sources of biological order at multiple spatiotemporal scales including evolutionary processes (Kauffman, 1993). In view of these issues, OPB represents thermodynamic principles, laws, and properties because of their relevance to all physical structures and processes and as a resource for annotating bond graph models.

QUALITATIVE, DISCRETE CAUSAL METHODS

Quantitative computational modeling as supported by the above modeling platforms provides a "gold-standard" test of the suitability of a mechanistic hypothesis to represent and explain empirical observations and data.

However, these benefits come at great cost in terms of the depth, breadth, and precision of the required data and constraints and, importantly, the difficulty of casting mechanistic insights into testable hypotheses and computational models. Such issues are particularly problematic as the size and intricacy of models increase for models of large-scale genomic expression and of metabolic processes.

A major constraint to the benefits of quantitative modeling and analysis of biological and, especially, cell biological systems is the difficulty of defining and acquiring sufficiently complete and accurate quantitative values to warrant a complete quantitative analysis. A major downside to quantitative analysis of complex biophysical networks is the exponential costs of defining the values of initial and outcome variables, as well as the often more difficult specifying and evaluating the constitutive properties in the network. A number of groups have, thus, explored alternative analytic approaches based on probabilistic modeling or qualitative causal reasoning.

For example, "Reverse Causal Reasoning" yields mechanistic insights into the interpretation of gene expression profiling data that are distinct from and complementary to the results of analyses using ontology or pathway gene sets (Catlett et al., 2013). This reverse engineering algorithm provides an evidence-driven approach to the development of models of disease, drug action, and drug toxicity. This approach may apply to datasets and physiological mechanisms other than gene expression systems.

Agent-Based Models

An agent-based model (ABM) consists of logical rules that apply to sets of autonomous physical "agents" as they interact in a spatial region. For example, one could use ABM to model (1) *molecules* that diffuse and encounter other molecules with whom they may, or may not, chemically react into product molecules, (2) *cells* that flow or diffuse through a portion of blood and may activate, inactivate, or destroy other cells on contact, or (3) *organisms* infected with a disease agent that may infect others on physical contact while migrating through physical space.

ABMs afford a high-level view of the populations dynamics of interacting entities by abstracting and simplifying agent–agent interactions to reveal emergent dynamical behaviors of populations of coupled agents. In both continuum and lumped-parameter modeling, the participating biological entities persist in time; none are created, and none are destroyed. This is perfectly appropriate for modeling blood flow through vessels, muscles contracting, or chemical flows among portions of chemicals. However, many important physiological problems can best be understood and modeled if the participants can be created, destroyed, or, even, change entity type. The key distinction is whether there are a countable number of participating entities or if the number of entities is uncountable high.

ABMs are spatially *discrete* and temporally *discrete* and are particularly suited for representing populations or sets of countable things such as biological cells or macromolecules like genes. Each thing is represented as an individual that can be created, destroyed, or changed into a different kind of thing according to specific biology-inspired rules. For example, cells within a population of cells (such as a set of cancer cells) are created and destroyed at discrete moments in time. Sluka describes this analytic approach, including the Cell Behavior Ontology (Sluka et al., 2014); Husakova presents how agent-based modeling and a declarative Agent Modeling Language could be applied to immunology (Husakova, 2015).

The Cell Behavior Ontology represents ABM modeling in which, for example, individual cells (Figure 5.7) of a particular type are represented, counted, and tracked as independent "agents" that are created, interact physiologically, and perish according to particular rule sets (An et al., 2009). For example, rather than modeling capillary blood flow as some sort of pseudo fluid, agent-based modeling represents each of the following cells as an independent agent with its own physical characteristics (size, shape, surface location) as it is driven by fluid forces and contact with other cell agents.

Cause–Effect Inference

The value of quantitative dynamical modeling rests in its use for testing and validating biophysical hypotheses. Model parameter values can be optimized to identify the best fit of the model to available data. However, these

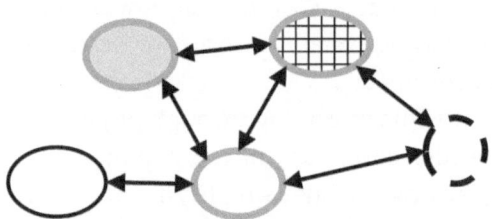

FIGURE 5.7 Topology of an ABM wherein a population of independent "agents" that have specific interactions on an individual basis with different kinds of agents to simulate, for example, molecular processes within a cell, or cell-to-cell interactions.

are computationally costly procedures that may result in near-optimum, but different, parameter sets that each sufficiently reproduce a dataset. However, quantitative inferences depend on identifying and observing the quantitative effects on one or more model state properties by small perturbations of the values of other system properties and parameters. This is a simple task for testing a small number of values but can be unwieldy and computationally expensive for large or highly-coupled models.

A low-cost, but informative, alternative is to perform qualitative perturbation analyses that mimic the sort of qualitative reasoning that when one is confronted with a system for which there is little or no quantitative data to use as constraints. Investigators routinely invoke verbal descriptions and reasoning such as, "If property A increases, then property B decreases." The appeal of binary, qualitative modeling is that it is intuitively simple but may expose critical, but unexpected feedback loops in complex models.

The limitations of binary (e.g., true/false, +/–) logics for representing ambiguous quantitative computations are immediately obvious. For mathematical *summation* or *product* relations between two qualitative values (e.g., + or –) are indeterminate with qualitative reasoning. Similarly, indeterminate is the polarity of a feedback loop. However, qualitative reasoning can identify cause–effect pathways.

Logical methods are used to model qualitative dependencies between system variables and system outcomes. Sometimes, models are coded or instantiated by a set of logical propositions that describe system composition and function. For example, a two- or three-dimensional "truth table" can represent which of a set of molecular components can combine into two or more macromolecular complexes. More implicit encodings are "state-transition" tables whereby the all-or-none expression of gene products (e.g., proteins, pathways) triggers the all-or-none activation of particular genes, combinations of genes, or biological traits.

The advantages of such binary logical methods are that they are easily and quickly computable and can handle large, easily-stored datasets such as gene activation profiles or ligand-binding maps. The limitation of declarative logics for biophysical modeling is that all-or-none behaviors are drastic simplifications and poor approximations for graded responses seen in (real) ligand-binding, gene-activation, and metabolic phenomena. As will be discussed in more detail in Chapters 5 and 6, we have adapted and extended logical methods to make a richer set of logical inferences including activation and inhibition, as well as revealing positive and negative feedback loops in complex physiological systems.

Chalkboard Semantics-Based Modeling and Causal Reasoning

We have developed Chalkboard, a Java application for modeling cell physiological systems based on a graphical, node-and-link modeling environment (Cook et al., 2001, 2007, 2013). Chalkboard offers a menu (at the left of Figure 5.8) of *Object icons* to represent cells, cell membranes, molecules and molecular parts, and compartments. The *Action* tool is a "smart" linker for creating functional connections between icons such as *binding* action (dashed line) between *Bind sites* or a phosphorylation action between a *Kinase* site and a *P-site*.

Chalkboard systems diagramming is based on a semantic structure that anticipates the formal ontological structures of OPB. Its representational vocabulary consists of three major classes (Cook et al., 2001):

1. *Entity* instances (2D icons) represent basic cell biological structures such as *Compartments* (e.g., intracellular space, intranuclear space), *Molecules* (e.g., protein, metabolite, ion, or polynucleotide), and *Functional sites* (e.g., molecular binding sites, catalytic sites). Each

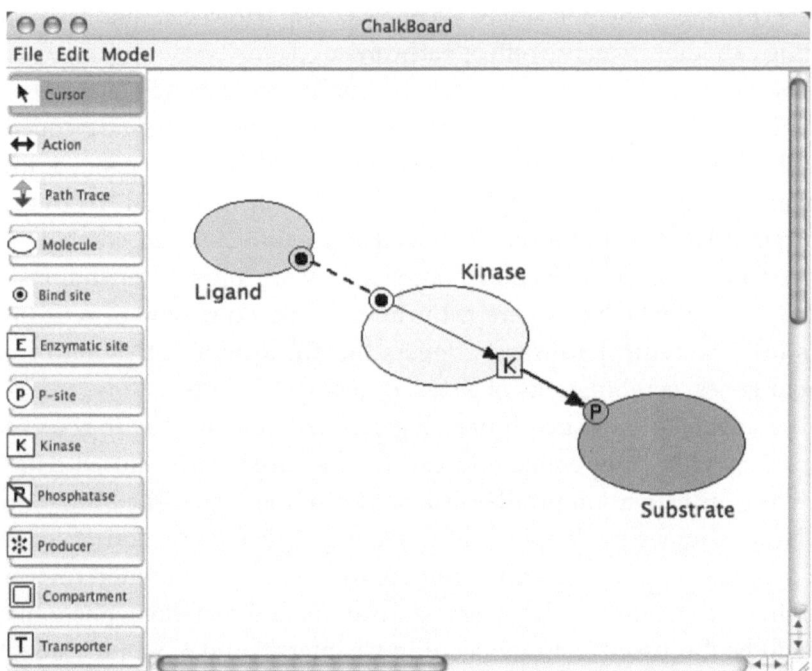

FIGURE 5.8 Chalkboard semantic modeling software implemented a hierarchy of graphical modeling classes to model basic biochemical entities and processes, and to support qualitative cause–effect reasoning.

Entity instance embeds a binary state variable (0 or 1) that represents an *amount* or *activity* a property of the entity.

2. *Action* instances (arrows) represent functional interactions by which the amounts and activities of *Entity* instances change the functional states of other *Entity* instances. For example, chemical reactions, transporters, or receptor–ligand binding reactions (represented as node–node linking arrows) represent material fluxes. Actions can be modulated (i.e., *activated* or *inhibited*) in a binary fashion as pathways by which entities can control other actions.

3. Each instance of an *Entity* icon or *Action* arrow embeds a binary (i.e., 0, 1) "physical property" that represents the *amount* of an entity or the *rate* of an action. For example, *Chemical flows* arrows represent a variety of chemical processes such as *Bind reactions* (dimerization) and *Transporter flow* (across a *Compartment* boundary). *Actions* can be modulated (e.g., activated or inhibited) to represent the complex cell signaling logic.

Chalkboard implements universally used, but informal, biophysical "causal propagation analysis", as a form of physics-based reasoning whereby events are propagated through a functional network according to embedded logical rules. For example, one might state that "an increase of plasma glucose concentration stimulates insulin secretion into systemic blood" and that "an increase of insulin level in systemic blood increases the rate of glucose uptake by fat cells". We have described and implemented a formalization of this approach in a prototype Java computer program, "Chalkboard".

The Chalkboard object model, that foreshadows the OPB ontology, is based on an inheritance hierarchy of biological "objects" and "processes" that represent real-life objects and actions using an inheritance hierarchy of graphical *icons* that represent biological structural entities (e.g., *cellular compartments*, *molecules*, and *functional sites*) and *arrows* to represent biological actions in which they participate such as molecular binding, enzyme catalysis, and phosphorylation. Entities and processes have physical properties such as the amount of metabolite or ligand, the volume of a cell cytoplasmic compartment, or a reaction rate or affinity of a biochemical reaction or of ligand binding.

We have developed software (see Chapter 7) that implements this analytical paradigm albeit by leveraging the automatic reasoning (Neal et al., 2016) based on of the Web Ontology Language (OWL).

Advantages and Limitations of Qualitative Methods

Qualitative methods such as those implemented by Chalkboard are attractive because they can be used to quickly model complex systems especially at early stages of when one is exploring and explicating particular hypotheses. Applications like Chalkboard (see others below) provide very facile tools to rapidly brainstorm and reason about system composition and functional interactions. Causal reasoning, as implemented by Chalkboard, can be a powerful tool for testing and visualizing causal pathways in an all-or-none manner to detect and display both local and distant, perhaps unexpected, effects of property perturbations on others in the system.

QUANTITATIVE SYSTEMS ANALYSIS

Quantitative systems analysis is the application of quantitative, physics-based methods to represent and analyze the biophysical bases of hypotheses about biophysical systems. Given the costs and rigors of acquiring and managing quantitative data about large-scale dynamical systems, these methods offer major computational challenges but also large rewards in terms of scientific insight and practical utility.

A biophysical model represents a set of causal relations by which physical continuants change according to the physical processes in which they participate. Such relations are directly represented in discrete-state models wherein the change in continuant states maps directly to the states of other continuants by causal processes that are themselves represented discreetly. Thus, the causal structure of such a model is fully represented in an all-or-none fashion such that each set of mappings constitutes a different causal hypothesis. Alternatively, ODE and PDE models express causal relations and biophysical states in a manner that depends, for example, on the quantitative property values of initial conditions, sizes or masses of continuants, rate constants of processes, or feedback gains, etc. Thus, the computational code of each quantitative model represents a spectrum of model instantiation according to choices of parametric values. This presents considerable complexity to the task of model validation, interpretation, and use beyond the discrete cause–effect.

For example, only a certain range of blood vessel caliber will be suitable for a blood flow model – a low value will under-represent observed flow rates while a too high value will overestimate flow rates. In either case, the model will not properly represent how vessel caliber determines blood flow rates and other cardiovascular outcomes. Whereas a discrete model can only represent a vessel as qualitatively open (e.g., as a 1) or closed (e.g., as a 0), a quantitative model demands a quantitative value for each parameter.

For example, a simple model of blood flow through a single capillary might represent the blood flow rate based on a model equation having parameters for length, inner diameter, and blood viscosity. Whether the equation is a sufficient model for capillary blood flow depends on how closely the model equations replicate observed data. Some models may be rejected simply on the basis of qualitative criteria – the model predicts that a computed property value decreases when the data show an increase or that model values oscillate when no such behavior is observed for the real system.

System Dynamic Modeling – Participants, Processes, and Properties

Investigators model and analyze systems for many reasons. One may need to test specific hypotheses about system function. Another may need a computational model to guide therapeutic decisions such as for vascular or orthopedic surgery. They may be motivated by simple curiosity or the need to explain empirical observations. What happens if this input is changed? What if a participant is disabled? How important is a given subsystem to overall system performance? How can a specific system output be increased? How can the system be isolated from environmental impacts? How sensitive is system performance to changes of system property values? Such questions may be answered with exhaustive empirical testing and observations; however, a validated computational model may answer many more such questions and do so at a fraction of the cost.

Key steps for developing, validating, and interrogating a computational model are to (1) identify the physical *participants* and *processes* that constitute the system, (2) specify those physical properties whose values are to be accounted for by the model, and (3) represent as mathematical or logical statements those causal dependencies by which property values depend upon one another. For example, a model of the first step in the glycolytic pathway could: (1) identify the pools of cytoplasmic of glucose, ADP, ATP, and enzymes that participate in the process of glucose phosphorylation, (2) specify the rate of glucose phosphorylation as an observed value to be explained by the model, and (3) derive mathematical equations to account for conservation of mass and for the rates of the various chemical reactions.

Modeling Continuum Entities – Partial Differential Equations, PDEs

PDEs are used to model and predict spatially continuous flows of fluids (e.g., blood or air) and mechanical stresses and strains of solids (e.g., muscle or bone). Such studies attempt to predict and explain how, say, fluid flow gradients and vectors as a function space in a flow field within a blood flow turbulence affects atherosclerotic plaque formation or how spatially-local bone pathologies affect bone stresses and bone growth. Continuum analyses of muscle and bone seek to explain the mechanisms of fracture and healing. Such methods, however, require substantial data and computational expense and are reserved for only a few applications. Continuum physics is yet to be represented by the OPB. Continuum dynamic modeling has been particularly advanced for the study of cardiac mechanics (e.g., see Nickerson et al., 2016; Wang et al., 2015) but is beyond the scope of this publication.

The applicable physical principles are the same as for discrete modeling except imagine that each voxel is modeled as a discrete entity that exchanges force, momenta, charge, and mass flows with its immediate neighbors. This forms, in effect, a network of spatially adjacent dynamical entities that obey the same mathematics as applied to discrete, lumped-parameter models except continuous entities are discretized into transform from discrete representation into continuum representation by conceptually shrinking voxel size until they converge to zero in all spatial axes in a mathematical process akin to transforming discrete differentials ($\Delta f / \Delta x$) into continuum derivatives (df/dx) where f represents a continuum spatial function, called a "basis function", of some physical property (e.g., velocity, pressure, temperature).

Modeling Systems of Discrete Entities – Ordinary Differential Equations, ODEs

Quantitative differential equations, such as above, are routinely solved using simulation software tools that compute and display the results of virtual "experiments". For example, one can readily compute and display the time course of V by providing an "initial" value (e.g., V_0) for V and values for each parameter (e.g., k and R, as above). Such are the enterprises and methods described in Chapter 5 which have burgeoned into a tower of babel of different analytical languages, methods, and platforms that substantially inhibit model and knowledge reproducibility and reuse.

To make modeling a bit more apparent, we have drafted a "pseudocode" (i.e., not really executable) version of a simulation model based on simple ODEs as implemented in various model analysis tools (see the box). This code is designed to simulate blood flow from a left ventricle, through the aortic valve, and into the aorta during a single heartbeat, during which the left ventricle contracts and pressurizes blood until the aortic valve opens and allows flow into the aorta.

```
// declare and annotate model variables
        Pv;              // pressure of blood in left ventricular
        Pa;              // pressure of blood in aorta
        Vv;              // volume of blood in the left ventricle
        Va;              // volume of blood in the aorta
        Cv;              // fluid compliance of left ventricular
                            lumen
        Rav;             // fluid flow resistance through
                            aortic valve
        Fva;             // blood flow rate through aortic valve

// set time control values for the period of simulation:
        t      = __;     // time
        tstart = __;     // start time
        tend   = __;     // end time
        delt   = __;     // time step

// loop on time steps for the duration of the simulation
        t = 0;                       // initialize simulation
                                        time
        Vv = Vv(0), Pv = Vv / Cv     // initialize ventricular
                                        volume, and pressure


                                     // begin simulation
while (Pv > Pa) (                    // establish a
                                        computational
                                        time-loop
                                     // for as long as Pv >
                                        Pa; i.e. there is
                                        flow
        Pv = Vv / Cv                 // calculate pressures
                                        per compliances
        Pa = Va / Ca
        Fva = (Pv - Pa) / Rva        // calculate aortic
                                        valve flow rate
        Vv = Vv - delt * Fva         // update volumes per
                                        flow rates
        Va = Va + delt * Fva
        t = t + delt                 // increment time by
                                        one step

        )
plot: Vv, Va vs. time                // display results
```

Network Thermodynamics and Bond-Graph Theory

Network thermodynamics was developed in the 1970s for the analysis of biophysical systems based on the storage and exchange of thermodynamic energy among participants in a dynamical system (Oster et al., 1973). Network thermodynamics is built on the solid foundations of conservation of energy, which gives it scope for representing and integrating, within a single analytical framework, multidomain biophysical systems such as biochemical and biomechanical systems (Le Rolle et al., 2005; Lefèvre et al., 1999; Margolis, 1985).

Despite the reliance on an unfamiliar and challenging bond-graph notation, recent results demonstrate the power and utility of this approach (Crampin et al., 2004; Gawthrop & Crampin, 2018).

The underlying theory and analytical advantage of the bond-graph approach lie in the generality of thermodynamic theory that is the underpinnings of all physical theory and the basis for describing, analyzing, and interrelating all physical phenomena across physical domains and spatial scales. Thermodynamics has been used to error-check biophysical models and calculations as a *post-hoc* error check to confirm that no stricture of conservation of energy has been violated. Network thermodynamics inverts this approach to base biophysical network models by assuming and constraining kinetic models to conform to energy conservation rules.

Network thermodynamics and its implementation as energy bond-graph modeling is an analytical framework for dynamical systems modeling. Almost simultaneously, Oster and Perelson developed methods for analyzing complex chemical reaction networks as networks of quantities of chemical potential energy and the flows of energy between chemical reaction participants (Oster et al., 1971, 1973; Perelson, 1975). This approach has been adapted and generalized for analyzing multidomain dynamical engineering systems (Karnopp, 1979; Margolis, 1985) and, more recently, applied to cardiovascular systems (Le Rolle et al., 2005; Lefèvre et al., 1999) and to biochemical pathways and networks as energy bond-graph theory (Gawthrop & Crampin, 2014, 2018). Bond-graph modeling is based on the premise that an energy-based analysis is fundamentally scale- and domain-free such that a single mathematical formalism can represent and compute across all manner of multidomain, multiscale dynamical systems fully and quantitatively based on the stocks and flows of thermodynamic energy.

Hybrid Qualitative, Quantitative Modeling

We have briefly described a variety of physics-based modeling methods that are applied to particular problems. However, solving some multi-domain problems requires a hybrid of analytical methods that are each suitable for a particular domain. For example, the Karr Laboratory has focused on implementing and developing WholeCellKB as hybrid qualitative and quantitative computational tools for simulation and analysis of a mix of cell process submodels (Karr et al., 2013) (see Figure 5.9).

Modularization – A Solution for Biophysical Modeling?

Whatever the analytical theory, a key issue for any modeling study is to determine the level of analytical granularity – does the analysis comprehend only organ-level entities such as hearts, vessels, and blood, or will the model be expressed in terms of cells and molecules, or some combination of these. Is it sufficient for analytical needs that the scientific questions can be satisfactorily addressed by modeling only biochemical processes, or only cellular processes, or only organ-level processes? Or, do the questions demand a multiscale approach involving physical processes that have molecules, cells, and organs as participants? If the latter, an effective modeling approach is to designate functional "modules" that encapsulate various processes by structural level (big or small), domain (chemical or electrical), or by temporal scope (fast or slow).

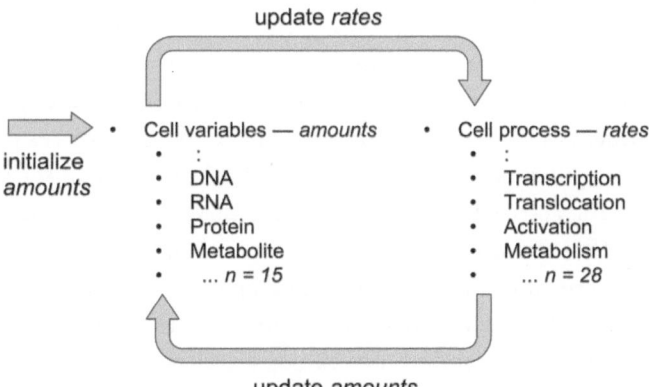

FIGURE 5.9 The WholeCellKB multimodel system consists of a partitioning cell variable (left) representing the physical states of various cell continuants, and domain-specific computational cell processes models that depend on and change the physical states.

The modularization approach is very common and very valuable for all manner of technical tasks from physical analysis to computer programming where the "black-box" approach is advocated for "object-oriented programming" (OOP) computer methods. Thus, an OOP programming object is a coded module that hides its internal computational processes behind a sparse interface so that other objects can address and use objects without needing to know or access knowledge of the object's internal states and processes.

Whereas this has demonstrated distinct advantages for programming systems designed by human, it has critical limitations for representing analyzing systems that are as intricately and highly-coupled as multiscale/multidomain biophysical systems (Neal et al., 2014). In this domain, it is quite unlikely that two investigators or projects can agree among many alternative (and, likely, conflicting) ways of parsing biophysical systems for a particular analytical need. Thus, we have recognized that modularization impacts and limits our ability to find and reuse modeling code from different sources.

"Goodness of Fit" – The Predictive Value of a Model

Whether a given model or its alternatives offer satisfactory explanations for an observed dataset is determined by quantitatively estimating how closely model outcomes match key observations based on defined quantitative datasets. For example, how do steady values for flow and state properties match observations? How do dynamic rates of decay or frequency match observations? Modelers use various statistical criteria to assess the so-called "goodness of fit" of model outputs as compared to observed data values.

The goodness of fit describes statistical criteria as measures of how well the model output values replicate a specific empirical observation that inevitably, being attributes of complex biological systems, include random components that must be accounted for. This requires establishing measures of goodness of fit that represent discrepancies between observed values and predicted model values. For example, one could simply subtract values predicted by the model from values actually observed. For example, there may be a single key datum that the model is intended to predict such as the increment of body weight at the end of a particular dietary restriction. If so, then a useful goodness-of-fit criterion is the simple difference between observed and computed values of weight.

As a more challenging example, consider comparing observed and model-predicted time courses of arterial blood pressure during a heartbeat.

To compute a prediction-to-observed error for the model, one might simply calculate the sum of all algebraic differences for all time points during a heartbeat. This would be satisfactory if pressures were monotonically increased (or decreased) during the entire heartbeat interval. However, it may turn out that the sum of systolic increments could be offset by decrements of pressure during diastole such that the net error is zero and, therefore, useless as a criterion for discriminating models.

The solution long used in statistical science is to define the overall error as the sum of the squares of the error at each time point – the "sum of squared errors" – such that positive and negative errors contribute equally to the net error computation. This is a very simple example of other error computational strategies that have been employed. For example, it may be determined that errors in predicting systolic pressures are far more critical for purposes of model development than predicting systolic pressures are far more critical than diastolic pressures so that they are weighted positively in the net-error computation. Defining a proper goodness-of-fit criterion is not easy and cannot be done by rote. Defining an error measure depends on the availability of datasets, and the hypothesis to be modeled and tested.

Sensitivity Analysis and Model "Optimization"

One key advantage of accurate and validated models of complex systems is that they can provide key insights into how specific processes and participants contribute to system function and operational outcomes. For example, although it may be intuitive that increasing myocardial contractile strength will increase cardiac output, this would be very difficult to demonstrate empirically for lack of methods to adjust contractile strength in the laboratory. One's intuition could, however, be satisfied, but not proven, by using a model to observe virtual increases in contractile strength in a model.

For another example, one might ask what level of a drug causes myocardial infarction or triggers rampant hyperglycemia? Is there a dynamical threshold for unstable oscillations of a system such as those underlying ventricular fibrillation? Such questions must be approached empirically or clinically with considerable caution as mistakes may be fatal. Whereas clear limitations and cautions apply to any modeling approach, quantitative methods afford deeper insights into system operations, functions, and outcomes. One key benefit to quantitative methods is that one can explore the so-called dynamic space of a system by learning how it responds to various parametric and boundary conditions. Is the system dynamically stable or for some parameters and conditions does it oscillate or exponentially approach a stable state within a range of normal operating parameters and outcome values? Such important questions cannot be realistically asked of qualitative models but are susceptible to quantitative analysis if carefully and rigorously performed.

SUMMARY

We have presented a necessarily brief overview of the various methods and assumptions that have been developed and employed for the analysis and understanding of multiscale, multidomain biophysical systems. Just as the sciences of multiscale biological structure have benefited from the advent of ontologies such as the Foundational Model of Anatomy, we have sought a comparable ontological approach for annotating, analyzing, and reusing physics-based analytical and modeling tools. In this next chapter, we introduce the Ontology of Physics for Biology based on a semantic approach to representing theories of classical physics.

CHAPTER 6

Ontology of Physics for Biology

T HIS CHAPTER BUILDS ON and formalizes the practices and theory of system dynamics as applied to the observation and analysis of complex, multiscale, and multidomain biological systems. We describe the ontological foundations and implementation of the Ontology of Physics for Biology (OPB) as a formal, semantic representation of physical entities, processes, quantities, and observable properties. It includes the spatiotemporal relations that extend available upper-level ontologies and classes that represent the network of the values of physical properties and the classical physical laws by which those values depend on one another.

OPB – AIMS, SCOPE, AND STATUS

OPB is a work in progress that has been curated with two aims. First, as an intellectual task, we have aimed to recognize and represent the basics of classical physics as articulated by engineering system dynamics, biophysics, and physiology. Second, we have aimed to provide the ontological tools to represent and annotate the physical content of available datasets, analyses, and models. We have developed the fundamentals of physics and engineering systems dynamics to serve both biomedical and the broader technical communities.

In Chapter 3, we described other ontologies that are particularly useful for annotating continuants (e.g., FMA, ChEBI, GO:cellular component), processes (e.g., GO:biological process, SBO), and phenotypes (e.g., PATO);

DOI: 10.1201/9780429469961-6

yet, these views of biological continuants and processes are largely descriptive and are not based on principles of physics. As discussed in Chapter 5, physiologists, biophysicists, and bioengineers are much more concerned with measuring *values* of physical quantities and comparing values predicted by hypothetical models based on empirical rules and the laws of classical physics.

OPB is not intended to represent all of physical reality but focuses on the classical physics of organismal physiology and biophysics. Accordingly, OPB does not represent

- relativistic physics near the speed of light

- quantum physics of subatomic particles

- astrophysics of planets and stars

- optical physics of light and lenses

- psychophysics of visual, auditory, and sensory perception

- statistical mechanics of heat and molecular motions

- information processing of neural networks

OPB represents gravitational potential fields or electrical potential fields as thermodynamic portions of energy that may exist without spatial limits. Similarly, it is often convenient in physical analyses to assume that some processes occur in a perpetual steady state having no temporal bounds. As the OPB is an ontology of biophysical analysis and not one of "reality" we have taken leave, as have others, to relax some of the strictures of realist philosophy in service to scientific practice (Lord & Stevens, 2010).

OPB represents the theory of classical physics as developed on the observation, measurement, and predictions of physics in terms of observable physical properties of physical continuants and processes. This is evident in the models and computations described in Chapter 4 that are based exclusively on reproducing and analyzing the values of observable physical properties where the entities themselves, continuants and processes, are implicit or declared only by annotations.

In this chapter, we will describe the basic representational framework as formulated for the development, annotation, description, and sharing of analytical models. Although developed in the domain of biology, biophysics, and physiology, its underlying assumptions, form, and content are applicable for the analysis of non-biological engineering systems. This chapter is an overview of the approach, scope, and contents of the OPB limited to those OPB classes and relations as are required for representing and annotating the biophysical content of available system dynamics models.

Those who are interested in investigating the nuts-and-bolts of the details of class hierarchical structure and class relations are invited to browse and download the self-documenting OPB which is, at this writing, a midsize ontology consisting of ≈850 OWL classes and ≈50 OWL object properties. The latest version of the ontology is available on BioPortal (https://bioportal.bioontology.org/ontologies/OPB) and on GitHub (https://github.com/SemBioProcess/OPB).

Given the breadth and depth of classical physics theory and application, we consider the OPB to be a work in progress that bears further development into other biophysical domains particularly in, say, neural signal processing, genomic systems, and others.

BACKGROUND AND MOTIVATION
OPB – Precursors and Neighbors

System dynamics has been deeply and intuitively understood by many workers ranging from auto mechanics to biophysicists. However, its mathematical, computational, and ontological bases began in the mid-nineteenth century and continued to evolve with advances in computer technology. To frame our work, we first present approaches to the ontological representation of physical processes and then turn to the ontological representation of physics that are used to explain such processes.

Biophysics – Maxwell Lays the Dynamical Groundwork
James Clerk Maxwell in a note entitled "Notes on the Mathematical Classification of Physical Quantities" outlined foundational aspects of the physical sciences that are useful for classifying and organizing a computational view of physical reality (Maxwell, 1871):

- The *first part* of the growth of a physical science consists in the discovery of a system of quantities on which its phenomenon may be conceived to depend". In the OPB, such measurable quantities are represented by subclasses of OPB:*Physical property*.

- The *next stage* is the discovery of the mathematical form of the relations between these quantities". In the OPB, such mathematical relations are represented by subclasses of OPB:*Physical dependency* whose "*classification*...is founded on the mathematical or formal analogy of the different quantities and not on the matter to which they belong".

These key observations by Maxwell anticipated more recent analytical approaches that have informed the design and implementation of OPB which are as follows.

"Tetrahedron of State" – Property and Dependencies
In the early 1970s, Oster, Perelson, and Kalchalsky developed "network thermodynamics" as an analytical approach that applies, by analogies, across multiple physical domains (Oster et al., 1971). They specified four classes (Figure 6.1) of observable, physical attributes as *force, flow rate, amount*, and *momentum* and five analytical dependencies as a basic analytical framework that generalized across physical domains. A nearly identical parsing of variable types and their mathematical dependencies across dynamical domains has been offered by Karnopp *et al.* as applied to multidomain "mechanotronic" systems (Karnopp et al., 2005). These model system components and their interactions are visualized as a "tetrahedron of state" (Figure 6.1) for dynamical systems that represent four physical properties (force, amount, flow rate, momentum) of system participants. They then define five binary relationships by which the values of those properties depend upon one another during physical processes within a single dynamical domain.

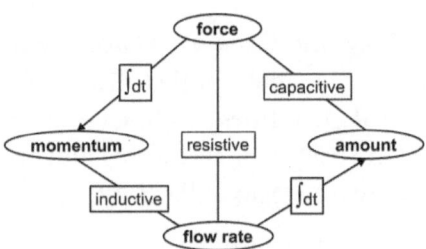

FIGURE 6.1 Karnopp's "tetrahedron of state" represents the four cardinal dynamical properties (force, momentum, flow rate, amount) that depend upon one another by constitutive laws (inductive, resistive, capacitive) and calculus laws relating flows to amounts and forces to momentum. Redrawn from Karnopp et al. (2005).

PhysSys – An Ontology for Engineering System Dynamics

These same principles were more formally represented in the PhysSys ontology (Borst et al., 1995, 1997). PhysSys formalizes a representation of network dynamics conceptualization of the mathematical expression of engineering physics based on the system dynamical approach described above. It was implemented by specializations and extensions of several sub-ontologies wherein physical continuants were represented by "Component" subclasses, occurrences, and processes by "Process" subclasses, with physical laws represented as mathematical statements derived from the EngMath ontology of engineering mathematics (Gruber & Olsen, 1994). Although PhysSys is no longer supported, its approach has informed the development of the OPB.

PhysSys aims to represent and analyze multidomain engineering designs and devices that span a limited range of engineering domains that include electrical, mechanical, and hydraulic domains as in Table 6.1.

PhysSys and other approaches to engineering analysis are well-adapted for representing and computing on problems within engineering domains. However, multiscale, multidomain biophysical systems present special challenges for defining and computing on their parts and processes. Models of such systems are necessarily tentative and incomplete because important cause–effect pathways may be unknown or have properties whose values are experimentally inaccessible or statistically uncertain.

Chalkboard – A Nascent Semantics of Biophysics

In Chapter 5, we introduced the Chalkboard application for the representation and qualitative analysis of cell biological systems, which has been developed in the object-oriented Java programming language. It is based on a linguistic metaphor that replicates the language that investigators use for sharing and discussing emerging hypotheses about the structure and function of cell biological systems.

Thus, the Chalkboard programming "object model" (Figure 6.2) represents physical objects like cells and molecules as "nouns" objects (i.e., nounModel.java) and physical processes as "verbs" objects (i.e., verbModel.

TABLE 6.1   PhySys Represents Three Property Classes (Stuff, Flow, Effort) across Three Engineering Dynamical Domains

Domain	Stuff	Flow	Effort
Electrical	Charge	Current	Voltage
Mechanical	Location	Velocity	Force
Hydraulical	Volume	Volume flow	Pressure

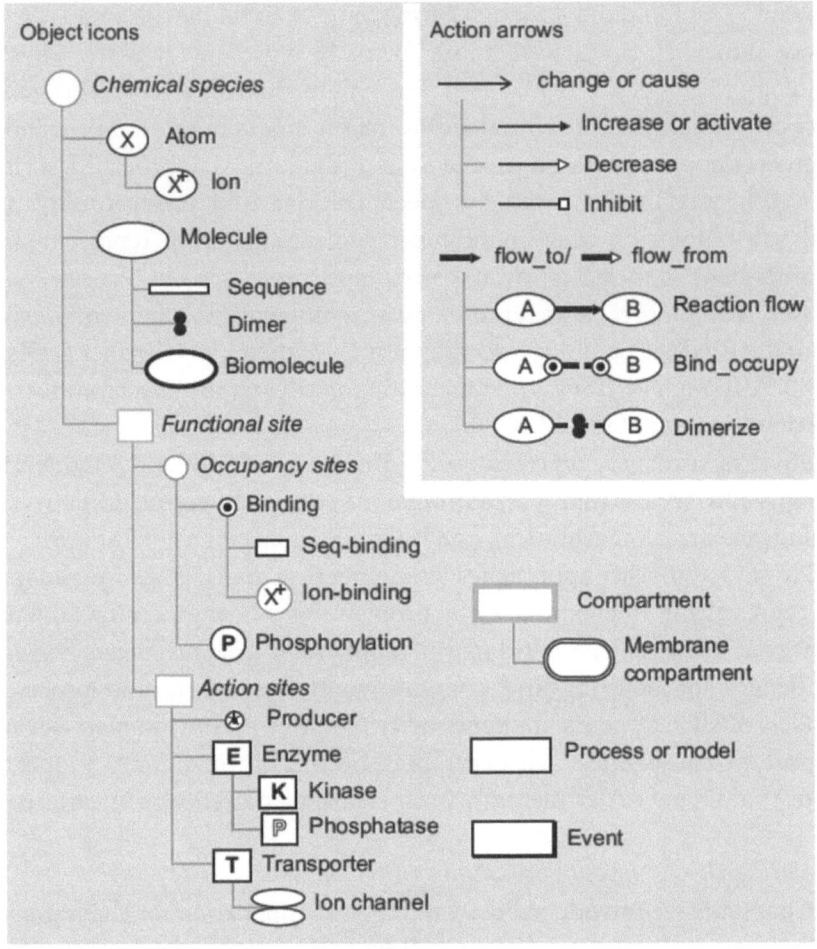

FIGURE 6.2 Chalkboard is based on a class hierarchy of graphical icons that represent cell biological entities (as Object icons) linked by processes (as Action arrows) to model complex cell processes.

java). Modeling is based on two inheritance hierarchies of graphical diagramming objects: (1) "Object icons" represent *continuants* such as cells, molecules, and their parts (e.g., binding sites, catalytic sites), and (2) "Action arrows" represent *occurrences* such as the flow of reactants through a chemical reaction (e.g., a Reaction flow), or a modulatory effect on such a flow or on some other modulatory action. Within its cell physiology modeling domain, Chalkboard computes qualitative cause–effect relations by tracking and displaying user-commanded increments and decrements of the (implicit) *amount* property of *Object icons*, or the *flow rate* property of *Action arrows* (see below).

Complex cell physiological models can be rendered using object icons linked by action arrows that represent and implement qualitative cause–effect relations such as the Chalkboard model shown in Figure 6.3.

Chalkboard is based on a small ontology of cellular structures (Object icons; a la FMA) and introduced a corresponding ontology of processes (action arrows) that presage their more comprehensive representation in OPB. Such a representation is based on cross-domain analogies that leverage both model representation and analysis.

Inspired by the efforts cited above, we sought to develop an ontology of classical physics that is useful for representing and computing on biophysical systems using modern ontology and reasoning methods. The result is the OPB that we have implemented in OWL (see Chapter 3) using the Protege ontology editor. The vast majority of available biomedical ontologies (see Chapter 3) are focused on things such as biological continuants as in the Foundational Model of Anatomy, processes as in the

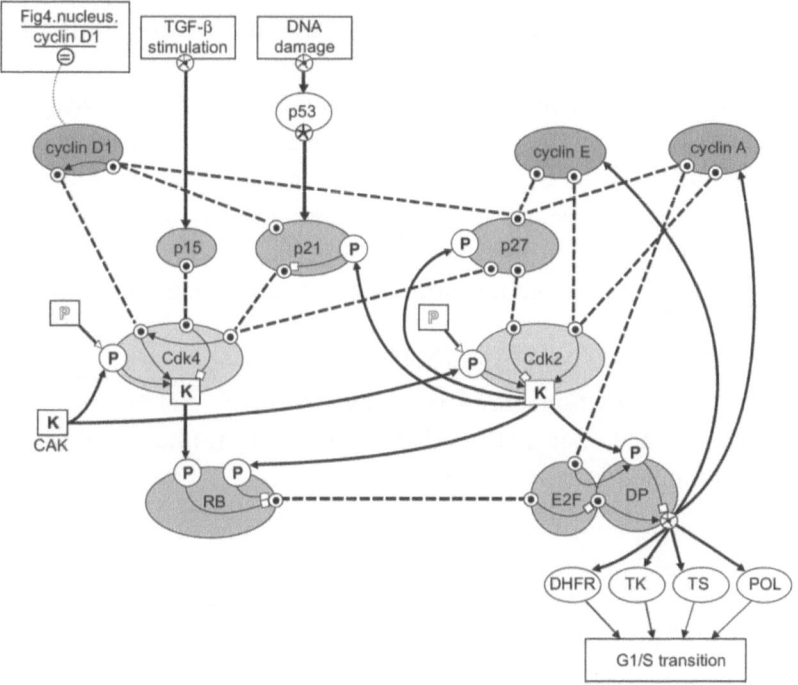

FIGURE 6.3 A Chalkboard model of the "G1S checkpoint" cell-cycle signaling system with participating molecules (e.g., "cyclin D1", "p15"), molecular functional sites (e.g., kinase sites "K", phosphorylation sites "P"), and the site–site arrows that represent processes (e.g., modulation or production as solid arrows, ligand-binding as dashed arrows).

Gene Ontology, and physical properties as terms and descriptors as in the Phenotypic Quality Ontology (see Chapter 3).

The OPB focuses on physical properties and the dependencies (e.g., laws) by which the values of the properties depend upon another as calculated in physics-based models and computations. Thus, we take a property-centric view that recognizes that a given model, say, of fluid flow is expressed entirely in terms of fluid flow rates, pressures, and resistance irrespective of the particular anatomical and physiological circumstance. Indeed, most physiological models can be readily repurposed because, in fact, variables and equations can be written generically wherein the particular biological applications are known only by their annotations.

OPB ORGANIZATION AND TOP CLASSES

The OPB top class, OPB:*Physics entity*, has two subclasses (Figure 6.4). First, OPB:*Physics real entity* classes represent what actually exists in the real world whereas OPB:*Physics annotation entity* classes represent those abstractions, models, and other representations that are crafted by the physical sciences to understand and model the real world.

This book focuses primarily on describing OPB:*Physics real entity* classes but we will start with the OPB:*Physics annotation entity* as a way to scope and motivate the aims and applications of the OPB in terms of annotating and analyzing the physics content of data sets, models, and literature. This is, necessarily, a brief introduction and superficial overview of the OPB to which readers are directed for more details and internal annotations.

OPB:PHYSICS ANNOTATION ENTITY

OPB:*Physics annotation entity* is a physics entity that is an analytical abstraction, concept, or method that is useful for annotating or analyzing a physical model.

- Physics entity
  - Physics real entity
    - Physics property
    - Physics continuant
    - Physics occurrent
  - Physics annotation entity
    - Physics domain
    - Physics model
    - Physics property attribute

FIGURE 6.4   OPB top class OPB:*Physics entity* has subclasses that distinguish what exists in the real world from the analytical tools used to analyze and model that reality.

OPB:Physics Domain

OPB is based on analogies among physical properties across physical domains by which, for example, the motion of a car along a road is analogous to the flow of water through a pipe or the flow of molecules through a biochemical process. These analogies are evident as parallel mathematical relationships between the values of physical properties across such domains. Ontologies of engineering system dynamics have relied on the organizing concept of engineering domains such as *mechanical, hydraulic,* and *electrical* domains each of which designates a model of interacting entities and processes describable by some particular kinds of mathematical relations. We have taken this approach and extended it to representing and annotating systems across the multiple dynamical domains that occur, are represented, and are analyzed by physiologists, biochemists, and biophysicists. Physics domain types are classified only as annotation entities because they apply to social constructs such as academic and scientific disciplines as much as they apply to the representation of certain kinds of properties, equations, and models.

In keeping with Maxwell's early suggestion, the OPB:*Physics domain* class hierarchy partitions the science of biophysical systems analysis into subfields for the purposes of annotating, subclassing, and interrelating OPB classes. In the remainder of this chapter, we will describe the major OPB classes that parse system dynamic theory into a representation that distinguishes: (1) the physical properties (OPB:*Physical properties*) that are the observable and computable attributes of biophysical systems and their processes, (2) the physical laws and quantitative relationships (OPB:*Physical dependency*) among the values of observable physical properties of (3) continuants (OPB:*Physics continuant*) that participate in (4) physical processes (OPB:*Physics process*).

The subclasses of OPB:*Physics domain* (Figure 6.5) are provided for high-level annotation and type-checking to assure compatibility and consistency of mathematical models, their computational components, and their empirical data resources. For example, it would be beneficial in searching for suitable models and datasets that are, for example, concerned with the biochemistry of islet physiology rather than blood circulation patterns among metabolic tissues. The concentration of a chemical species ought not to be taken as the fluid pressure of blood. The domain hierarchy is presented to illustrate and parse the extraordinary breadth required of the OPB as a representation of the analytical practices of bioengineers, biomodelers, and biophysicists.

- Physics domain
  - Dynamical domain
    - Material dynamical domain
      - Diffusion kinetic domain
      - Solid mechanical domain
      - Electrochemical domain
      - Fluid kinetic domain
      - Chemical kinetic domain
      - Cotransport domain
    - Immaterial dynamical domain
      - Heat kinetic domain
      - Electrical domain
      - Magnetic domain
      - Gravitational domain
  - Thermodynamical domain
  - Spatial domain

FIGURE 6.5   OPB:*Physics domain* classes represent various domains or disciplines of analytical investigation within the physical sciences including classes of OPB:*Dynamical domain*, OPB:*Thermodynamical domain*, and OPB:*Spatial domain*.

The OPB:*Dynamical domain* class (Figure 6.5) represents domains within which dynamical continuants exchange energy or information by the dynamical processes in which they participate. The OPB:*Material dynamical domain* and OPB:*Immaterial dynamical domain* subclasses apply, respectively, to continuants that are material (i.e., composed of atoms) or immaterial (e.g., electrical field). OPB:*Dynamical domain* classes are made available for annotating, type-checking, and interrelating instances of OPB dynamical classes for annotating and computing on instances of OPB:*Physical property*.

The hierarchy of dynamical domains further distinguishes OPB:*Immaterial dynamical domain* and OPB:*Material dynamical domain* depending upon whether what flows or moves is *immaterial* such as electrical charge or heat, or is a material such as a chemical, a fluid, or a solid. The OPB:*Transducer domain* encompasses continuants and processes that are of more than one domain. For example, a heart pumping blood is participating in both the fluid and solid domains, and its governing dependencies include physical properties from both domains. As other examples, a nerve transduces a mechanical stretch into the firing rate of action potential in its axon and the chemical kinetic transformation of metabolic molecules produces cellular energy in the form of adenosine triphosphate.

Thus, OPB:*Dynamical domain* classes are provided for use as terms for annotating and type-checking model instances including their data and code variables in physics-based models and analyses. This domain hierarchy partitions fields of engineering and physics investigation and practice. In fact, it can be argued that such domains may be closer to social or academic conventions than to real physical constructs encompassing continuants, processes, or properties. Whereas some datasets and models may encompass only biochemistry or only fluid dynamics, many can be multidomain. Such annotations can assure that, for example, continuants in one domain are not participating in processes defined for a different domain or that a variable annotated to one dynamical domain is used, inappropriately, in equations from other domains. The curatorial challenge in developing the OPB has been to develop a concise representational architecture that is as simple as possible but is capable of representing the basic computational approaches of biophysics and system dynamics.

The challenge of analyzing and simulating multidomain, multiscale dynamical systems is that it is usual that continuants and processes in one domain, such as biochemistry, are coupled to those in another domain such as the electrochemistry of ion channels or the molecular mechanics of muscle contraction. The challenge for comprehending, computing, and validating such multidomain models is that they comprise technologies, methods, and terminologies developed in domain-specific technical silos.

OPB:Physics Model

As part of the ontology itself, we provide English definitions. Thus,

> OPB:*Physics model* – A type of *Physics Annotation Entity* that is an analytical representation of one of more structurally or functionally related physics real entities.

A physics model may be used to annotate, encode, or express a biophysical theory, hypothesis, or explanation. As such, it shares a representational space with other biomedical and engineering ontologies (Cheong & Butscher, 2019; Collins et al., 2004; Erson & Cavusoglu, 2007; Gündel et al., 2013).

OPB:Physics Real Entity

OPB defines its foundational classes as follows:

> OPB:*Physics real entity* – A physics entity that is a continuant, an occurrent, or a property thereof that exists, or may exist, in the real world, and that can or may be observable by physical methods.

This is a purposely broad definition that aims to include most aspects of biophysical knowledge representation and functional analysis.

> OPB:*Physics continuant* – Physics continuants represent real entities that are portions of primitive physics entities and may be spatially-bounded (e.g, a heart, a portion of blood, a cell's cytoplasmic sodium ions) or may be spatially-unbounded (e.g., an electrical field or gravitational potential field).

The OPB:*Physics continuant* class has subclasses for OPB:*Spatial entity* that recapitulate OBO classes, but includes an OPB:*Dynamical entity* class that is defined as

> OPB:*Dynamical entity* – A physics continuant that occupies a spatial region is a bearer or conduit of energy and/or information, has dynamical properties, and can participate in a dynamical process.

These definitions lay the groundwork for defining instances of the class OPB:*Dynamical process*, itself a subclass of OPB:*Physics occurrent*, as

> OPB:*Dynamical process* – A physics occurrent that is the flow or exchange of matter, charge, or energy amongst dynamical entities that are participants in the process.

The main aim of OPB is to provide a knowledge resource for annotating, representing, computing, and reasoning about models of physics-based systems. Such models are cast in terms of variables, whether quantitative or qualitative, that represent the observable physical properties of the things and processes that they intend to model. Indeed, the things and processes themselves are, typically, declared and known by annotations and text descriptions of the model. Thus, whereas there are a plethora of biomedical ontologies devoted to representing biological continuants and processes, we have seen the need, as have others, to formally represent, first, the physical properties, themselves, as attributes of continuants and processes, and, second, to represent the laws and equations of physics that describe how the values of such properties depend upon one another.

OPB:*PHYSICS PROPERTY* – OBSERVABLE AND COMPUTABLE

The OPB:*Physics property* class is defined as "a physically observable attribute of a physics continuant or process that can be measured and represented as a scalar, vector or tensor, or as an aggregate of such measures, or as can be computationally derived from such measures." Such properties are the basis for empirical studies, statistical analysis, and model hypothesis expression and testing. Indeed, it is the properties of entities that are the basis for analysis across the gamut of statistical and dynamical model analyses.

The actual entities that bear the property are known and identified exclusively by naming convention or by annotation. For example, a physically observable attribute might be the fluid pressure of blood in a ventricle that is represented by a variable, P, in a generic equation for calculating the blood flow rate. Unless, "P" is explicitly annotated, its meaning is lost unless annotated by some naming convention (e.g., "PLV") or by some in-line text annotation (e.g., "pressure of blood in the left ventricle"). Although such a human-readable annotation is useful, a much better approach is to establish a format for machine-readable annotations such as the "composite annotation" format that will be discussed in the next chapter.

Dynamical Properties

OPB:*Dynamical property* (Figure 6.6) is defined as "A physics real entity that is an attribute of a physics continuant or occurrent that can be measured by physical means or can be computationally derived from such measures".

Instances of OPB:*Physical property* are those aspects of reality that can be observed, measured, and recorded as data in an essentially empirical approach based on observations using physical means such as a voltmeter (electrical potential), meter stick (length), balance scale (mass), or clock (time).

- Dynamical property
  - Dynamical rate property
    - Flow rate property
    - Force property
  - Dynamical state property
    - Momentum property
    - Amount property
  - Dynamical acceleration property
    - Electrical acceleration
    - Solid acceleration
    - Fluid acceleration
  - Physical state
    - Structural state
    - Dynamical state
    - Thermodynamical state
    - Process occurence state
    - Boundary state
    - Physical constant

FIGURE 6.6   Class hierarchy of OPB:*Dynamical property.*

The very core of dynamical analysis is to determine how the *rate* at which a dynamical system changes is based on the current dynamical *state* of the system. Even our kindergarten physics (Chapter 4) tells us that the amount of air in a lung at a moment in time is the net rate of air in or out of the lung cavity. At any instant in time (t) during inhalation, the faster the air flow rate (F) into the lungs, the more the rate of change of the total lung volume (V), and vice versa during exhalation. This dynamical relationship is captured quantitatively using the formalism of calculus whereby at any instant in time, i.e., $dV(t)/dt = F(t)$, or, inversely, $F(t) = V(t) + Vo$, where Vo is the value of volume at some initial instant of time; i.e., a baseline value.

As discussed in Chapter 5, the volume of air is a *state* property because it is a measure of the amount of air measured by volume at a moment in time. As the air is exhaled, the amount of air that flows out during a time interval is a *fluid flow rate* property (OPB:*Fluid flow rate*) whose value is a measure of the rate of change in the amount of air, OPB:*Fluid volume* of the air in the lung both of which are properties in the OPB:*Fluid kinetic* domain. This state/rate distinction is at the core of models' dynamical systems across multiple kinetic domains such as those below:

> *State* variables measure the physical *state* of an entity at a moment in time, for example:
>
> the *volume* of air in a lung (V, as above),
>
> the *length* of a contracting muscle, or
>
> the *amount* of charge separation across a cell membrane.

> *Rate* variables measure the *rate of flow* or *motion* at a moment in time, for example:
>
> the *flow rate* of air through the trachea (F, as above),
>
> the *velocity* of shortening of a muscle, or
>
> the *current* through ion channels in a membrane.

It is the goal and value of system dynamical analysis to understand, describe, and compute how the values of these various observable physical properties depend upon one another as the states of continuants are changed during the processes in which they participate. In the following, we will describe and classify such properties and the physical laws and constraints as discovered and used for analyzing physical systems.

OPB:*Dynamical Property* Classified by Dynamical Domain

The essence of system dynamic analysis is that it generalizes, by analogy, across multiple dynamical domains as summarized in the following chart and as described above. Note that momentum properties apply only to *material* flow in fluid and solid kinetic domains and to charge flow in the electrochemical domain (Figure 6.7).

OPB:*hasPhysicalDomain* Generalizes Physical Rules and Laws

As developed by OPB predecessors, the value of a general view of classical physics means that the continuants, occurrences, properties, and relations can be represented in a generalized form and then specialized to apply across the range of biophysical domains. For example, OPB represents one set of stock-and-flow rules and equations, such as conservation of matter, that apply across several physical domains (Figure 6.7). To assure that, for example, a fluid flow rate law applies only to fluid entities with fluid properties, OPB includes the OPB:*hasPhysicalDomain* property which is defined as a "topObjectProperty that relates a physical property, continuant, or occurrent to the physical domain to which it applies."

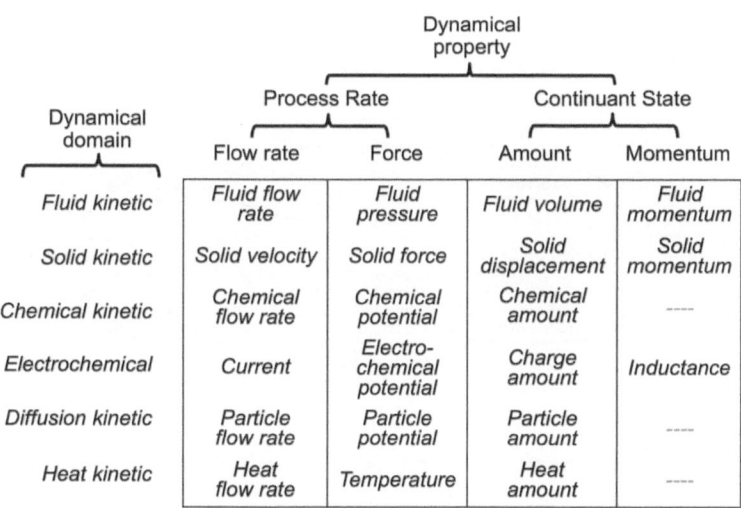

| Dynamical domain | Process Rate | | Continuant State | |
	Flow rate	Force	Amount	Momentum
Fluid kinetic	Fluid flow rate	Fluid pressure	Fluid volume	Fluid momentum
Solid kinetic	Solid velocity	Solid force	Solid displacement	Solid momentum
Chemical kinetic	Chemical flow rate	Chemical potential	Chemical amount	----
Electrochemical	Current	Electro-chemical potential	Charge amount	Inductance
Diffusion kinetic	Particle flow rate	Particle potential	Particle amount	----
Heat kinetic	Heat flow rate	Temperature	Heat amount	----

FIGURE 6.7   OPB:*Dynamical property* subclasses (such as OPB:*Fluid flow rate*) for selected domains that are subclasses of OPB:*Dynamical domain* (such as OPB:*Fluid kinetic domain*).

Dynamical Properties Are Attributes of Continuants and Processes

As discussed above, there is a fundamental ontological distinction between things, *continuants*, that exist at a moment in time, such as some blood or a bone, and *processes* that occur over a span of time during which participating continuants change either quantitatively (e.g., grow, move) or categorically (e.g., become a different kind of entity). Thus, continuants and processes are fundamentally different and have, as attributes, different instances of OPB:*Physical property*. Following the precedent of classical physics, OPB distinguishes (Figure 6.7) dynamical *rate* properties (force, flow rate) from dynamical *state* properties (amount, momentum) across the various dynamical domains. Figure 6.8 illustrates two important aspects of the OPB.

First, it diagrams how OPB:*Physical property* instances are attributed by the OPB:*hasProperty* relation to continuant instances and to process occurrences according to the schema shown in Figure 6.8. Second, it introduces dynamical dependency relations by which the values of properties depend on one another. Arrows represent instances of the OPB:*hasProperty* relation that is defined as "A topObjectProperty that relates an instance of a physical entity to an instance of a physical property that inheres in the entity."

Second, Figure 6.8 diagrammatically represents the temporal derivative and integral dependencies (subclasses of OPB:*Dynamical dependency*) by which, for example, the value of a rate property (e.g, flow rate or force) is

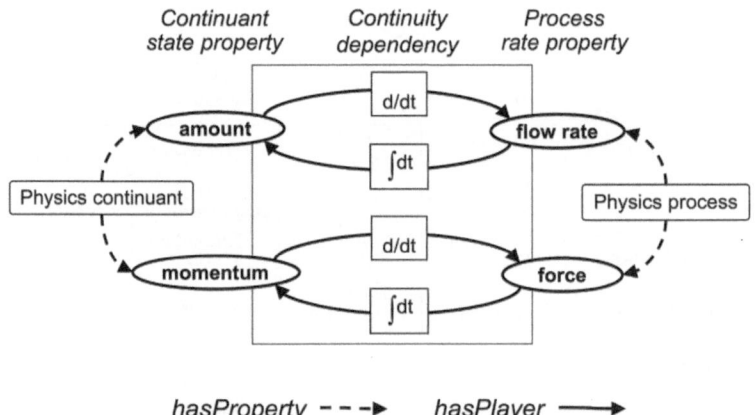

FIGURE 6.8 OPB:*Dynamical entity* subclasses, such as OPB:*Portion of fluid*, that have an instance of each of four OPB:*Dynamical property* classes (i.e., flow rate, amount, force, momentum).

the temporal derivative of a corresponding state property (e.g., amount or momentum), as described in the following.

OPB:*PHYSICS DEPENDENCY* – PHYSICAL LAWS AND CONSTRAINTS

The essence of system dynamic analysis is the specification of the flow rates of processes that affect the physical amount of the continuant stocks that comprise the system. In a dynamical network, one or more such flow processes link various entities that are sources and sinks for the flowing material. It is imperative, therefore, that the polarity of flows be correctly linked to the rates of change of the states of continuants. Differential equation analyses equate temporal rates of change of state variables (i.e., "left-hand sides") to the sum of flow rates (i.e., "right-hand sides") of "stuff" that are the processes.

OPB:*Physics dependency* is a subclass of OPB:*Physics occurrent* and is defined as "A physics occurrent that relates the existence and changes of attribute values of physics continuants to the occurrences and time courses of physics processes." It is so defined because a dependency, like a process, represents the *disposition* of one or more continuants to change in a particular way (Arp et al., 2015). Chemical reactants have a disposition to participate in a chemical reaction and do so in a particular way that observers describe in terms of particular physical laws and constraints.

Such laws and constraints are represented by occurrences of OPB:*Physics dependency* that represent the physics-based rules and laws invoked to analyze flows material, electrical charge, and energy among participants in physical processes. The class OPB:*Physics dependency* is defined as "A physics continuant that relates the existence of and changes of attribute values of a physics continuant to the occurrences and time courses of a physics process in which it participates."

Whereas *hasProperty* relates instances of properties to continuants or to occurrence of processes, the *hasPlayer* object property relates property instances to occurrences of OPB:*Physics dependency*.

OPB:hasPropertyPlayer Relations

Occurrences of OPB:*Property dependency* represent quantitative relationships between instances of OPB:*Physical property* that are instances of OPB:*hasPropertyPlayer* (Figure 6.9; that are subclasses of OWL:*topObjectProperty* in the OPB). For example, a computational expression of Ohm's law represents an instance of OPB:*ResistiveDependency*

- hasPropertyPlayer
  - hasConstitutivePlayer
    - hasNegPropertyPlayer
    - hasPosPropertyPlayer
  - hasStatePlayer
    - hasMomentumPlayer
    - hasAmountPlayer
  - hasThermodynamicPlayer
  - hasRatePlayer
    - hasForcePlayer
      - hasForceSinkPlayer
      - hasForceSourcePlayer
    - hasFlowPlayer
      - hasFlowSinkPlayer
      - hasFlowSourcePlayer

FIGURE 6.9 Subclasses of OPB:*hasPropertyPlayer* are relations between instances of OPB:*Physical property* and the occurrences of OPB:*Physical process.*

among OPB:*Physical property* instances that represent flow rate (i.e., I_{12}), a force (i.e., $V_1 - V_2$), and a constitutive parameter (i.e., R_{12}). OPB: *hasPropertyPlayer* relations link instances of OPB:*Physical property* to instances of OPB:*Physics dependency* across multiple domains (OPB:*Physics domain*). The *hasPropertyPlayer* relations can assure that a given dependency is expressed in terms of the correct kind of physical properties. Furthermore, the polarized player relations (e.g., *hasNegPropertyPlayer* or *hasFlowSinkPlayer*) represent how perturbations of property players relate qualitatively to each other as needed for qualitative reasoning as described in the next chapter.

As will be described in Chapter 7, these logical statements can be implemented as OWL logical restrictions that establish the basis for reasoning about how discrete changes in the values of property players depend upon one another. For example, the current flow rate is negatively affected by an increment in resistance while it is positively affected by an increment of flow-driving force property.

OPB:*Calculus Dependency* – Temporal and Spatial

Calculus, both derivative and integral, is absolutely fundamental to physical analysis along both temporal and spatial dimensions. We have discussed the fundamental dependencies that relate system states to their temporal rates of change represented by OPB:*Calculus dependency* classes. For example, the net flow rate of blood out of a ventricle directly diminishes the amount of blood in the ventricle, a relation expressed using the mathematics of calculus. Thus, as shown in Figures 6.10 and 6.11, *flow*

- Calculus dependency
  - Spatial calculus dependency
    - Spatial integrative dependency
    - Spatial derivative dependency
  - Temporal calculus dependency
    - Temporal integrative dependency
    - Temporal derivative dependency

FIGURE 6.10   OPB:*Calculus dependency* classes represent calculus-based mathematical relations whereby, for example, flow rates are temporal derivatives of amounts and, reciprocally, amounts are temporal integrals of flow rates.

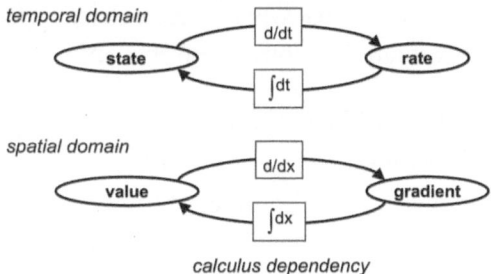

calculus dependency

FIGURE 6.11   Graphical mapping of property players (amount, flow rate, etc.) and their calculus dependencies (OPB:*Spatial calculus dependency*, OPB:*Temporal calculus dependency*) that represent conservation relations in the temporal and spatial domains.

rates are temporal derivatives of *amounts*, represented by OPB:*Temporal derivative* occurrences, and reciprocally, *amounts* are temporal integrals of *flow rates* represented by OPB:*Temporal integral* occurrences.

The calculus dependencies are the foundations of dynamical systems analysis but two other system dynamical dependencies, *conservation* dependencies are essential for modeling and understanding system behavior.

Instances of OPB:*Calculus dependency* represent the basic conservation laws for matter, electrical charge, energy, momentum, etc. whose magnitudes within a closed system can change only by exchange across the system boundary. These four kinds of physical stuff are thus conserved – i.e., cannot be created or destroyed – within the system. Therefore, the temporal derivative of an *amount* property of a portion of a conserved substance is quantitatively defined to be identical to the *flow rate* of the substance across the boundary of the portion of the substance.

OPB:*Calculus dependency* classes also represent occurrences of *continuity equations* that are known familiarly as conservation laws. Most simply stated, continuity equations represent processes wherein the *flow* of some conserved quantity – e.g., material, charge, energy, and momentum – into a closed space, changes the *amount* of that quantity in the space in proportion to the net *flow rate*. The result is a net increment of the *amount* of the quantity in addition to the amount initially in the space. The amount of a conserved species, X(t), at any moment in time (t) is expressed by the mathematical temporal integral of the flow rate of X, $F_x(t)$, during an interval that begins with an initial value X_0. Note that the continuity dependency for momentum (p) is simply a restatement of Newton's second law that is more familiarly as f=ma, wherein a force (f) applied to a mass (m) accelerates (a) to increment its momentum (p=mv).

OPB:Constitutive Dependency, OPB:Constitutive Property

Calculus dependencies hold for quantities of matter, charge, energy, and momentum defined by a spatial boundary but not on the specific material properties of the constituents such as their mechanical stiffness, fluid viscosity, or electrical resistivity that are the so-called *constitutive properties*.

OPB:*Mono-Constitutive Dependency* of a Single Participant

The class of constitutive dependencies and properties are attributes that are functions of the geometry and material composition of players in a process – how long and wide is the vessel, and how viscous is the blood. By contrast, constraint dependencies, as described above, depend only on the topology processes and their participants – how does the flow of substance across a boundary change the amount of substance within the boundary irrespective of the physical properties (e.g., viscosity or density) of the flowing substances.

Figure 6.12 illustrates these fundamental physical properties and constraint dependencies we use OPB class maps that amount to extension and transformation of Karnopp's "tetrahedron of state" (Figure 6.1) in which ovals represent instances of property values, and rectangles represent instances of conservation constraint dependencies, and arrows indicate the polarity of the dependencies. For example, a flow rate is the temporal derivative of a net amount, and a net change of amount during a period of time is a (indefinite) temporal integral of flow rate.

Constitutive dependencies comprise a pleomorphic and endless set of occurrences that are defined in terms of how processes occur in terms of the specific geometrical and material properties of process participants. The rate of blood flow through a vessel depends on the specific shape and size of the vessel and the viscosity of blood, which itself depends on the

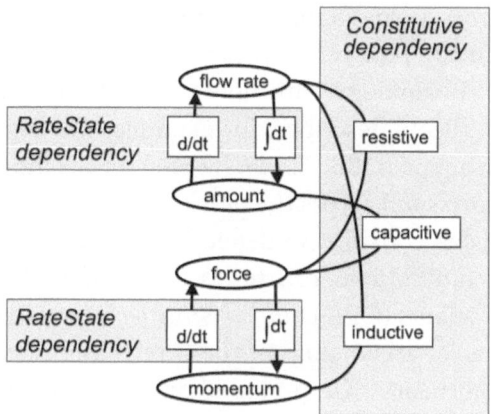

FIGURE 6.12   OPB:*Constitutive dependency* classes represent the classical laws of physics as dependencies among the values of instances of OPB:*Dynamical property* such as Ohm's law, Hooke's law, and the laws of induction.

blood flow rate. The contractile force developed by a muscle depends on the extent and rate of elongation, its length and shape, and the complex elastic properties of the muscle fibers that vary with elongation. The flow rate of blood through a vessel is determined by a *constitutive* dependency on vessel geometry and on the viscosity property of blood.

OPB represents nine subclasses of OPB:*Constitutive dependency* that account, to a substantial degree, for the broad range and depth of physiological and biophysical phenomena across functional domains as well as structural and temporal scales. Instances of OPB:*Constitutive dependency* represent quantitative biophysical relationships for processes whose activity depends on material properties, structural composition, and spatial features of the participants in the process. This is a diverse and pleomorphic class of dependencies that offers many representational challenges for discriminating specific occurrences that differ in a subtle way such as whether a given elastic element (like a tendon or muscle) has its force depends in a linear or nonlinear manner or whether such an elastic element includes a viscous, shock-absorbing component.

OPB:*Constitutive dependency* subclasses represent a broad array of biophysical rules and empirical laws that have been observed, analyzed, and modeled for various research and clinical purposes. OPB:*Constitutive dependency* subclasses are based on foundational laws of classical physics first expressed in the 18th century. For example, Ohm's Law is a kind of OPB:*Resistive dependency* by which the electrical potential difference

(OPB:*Voltage*) across a conductor is proportional to the electrical current flow rate (OPB:*Charge flow rate*) times an electrical resistance parameter (OPB:*Resistance*). By analogy with flow processes in other domains, the same construct usefully describes, for example, the dependence of fluid flow rate through a pipe (OPB:*Fluid flow rate*) on the fluid pressure difference (OPB:*Fluid pressure*) across the pipe.

OPB distinguishes constitutive dependencies that involve a single participating entity (OPB:*Mono constitutive dependency*) from those that involve two participants (OPB:*Dual constitutive dependency*).

Such dependencies are analogs of the three basic electrical circuit elements of electrical theory that are based on linear models that are classically defined as

> *Resistance* (R) is the ratio of electrical potential (a force) *across* the element to the *flow rate* of electrical charge (i) *through* the element; according to Ohm's law.

> *Capacitance* (C) is the ratio of electrical potential (a force) *across* the element to the *amount* of charge (q) *contained within* the element; analogous to Hooke's law.

> *Inductance* (I) is the ratio of electrical potential (a force) *across* the element to the *rate of change of flow rate* (di/dt) through the element; Henry's law.

Elementary circuit theory characterizes these dependencies as linear and time-independent "ideal" resistors, capacitors, and inductors for engineering use. Thus, OPB:*Constitutive dependency* classes are defined as more general analogs of electrical circuit elements as below.

OPB:Resistive Flow Dependency
Resistive flow dependency classes (Figure 6.13) represent the rate of flow of stuff and the attendant flow of energy where the flows are driven by a force differential between the flow's source and sink.

Subclasses apply to flows in other biophysical domains for which forces are defined. Thus, a resistive dependency is a generalization on Ohm's law that are force–flow relationships where the electrical current (I, a flow rate; OPB:*Flow rate property*) through an electrical conducting pathway depends linearly on the electrical potential difference (V, a force; OPB:*Force property*) across the conducting path. In cases, where V

- **Resistive flow dependency**
  - **Chemical resistive dependency**
  - **Electrical resistive dependency**
  - **Fluid resistive dependency**
  - **Solid resistive dependency**
  - **Heat transfer dependency**

FIGURE 6.13   OPB:*Resistive flow dependency*: A constitutive dependency by which a flow rate of a process depends on the difference in force properties of the process participants.

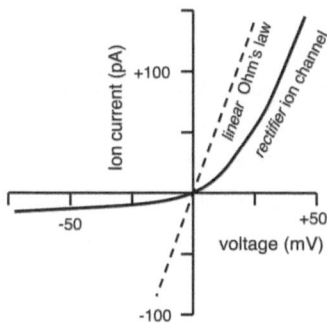

FIGURE 6.14   The distinctly nonlinear, non-Ohmic dependence of a "rectifier" ion current flow (pA) on membrane voltage (mV) is a critical aspect of membrane ion channel function.

and I are proportional according to the value of a resistance parameter (R, OPB:*Resistance*). Ohm's law can be written as $V=IR$, or as its inverse expression, $I=VG$ (where $V=$voltage, electrical potential across the element, $I=$electrical current flow rate through the element, $R=$resistance, a constitutive proportionality, $G=$conductance$=1/R$, which is also a constitutive proportionality).

Voltage-driven ion channel currents (Figure 6.14) can be distinctly nonlinear. For example, ions (e.g., K^+, Na^+, Ca^{++}, Cl^-) diffuse through membrane protein "ion channels" at rates that depend on the "driving force" that combines membrane voltage, and the electrical conductance of the channel for the particular ion. There are hundreds of ion channel types distinguished by their permeability to specific ions and are classed accordingly as potassium ion (K^+) channels, sodium ion channels (Na^+), calcium ion channels (Ca^{++}), and chloride ion channels (Cl^-) although none are strictly selective for a single ion type. The online database, Chemical Entities of Biological Interest (ChEBI; Degtyarenko et al., 2008) represents and classifies ion channels whose functions are well described by Hille (2001) and further discussed by Sakmann and Neher (1983).

Vascular blood flow. Similarly, blood flow through large arteries (e.g., aorta) is commonly modeled as a simple, proportional dependency of flow rate and pressure differential. However, as illustrated below, blood is not a simple, "Newtonian" fluid (like water) for which flow rates are linearly related to driving pressure forces.

The analogous flow dependency for chemical flow processes is the OPB: *Chemical resistive dependency* that represents the dependency of a chemical flow rate on the difference in the *chemical potentials* (OPB:*Chemical potential*) of its reactants. Thus, chemical resistive dependencies are analogous to ion flows driven by electrochemical potential (OPB: *Electrical potential* or OPB:*Electrodiffusional potential*) and fluid flows driven by fluid pressure (OPB:*Fluid pressure*). Computing chemical reaction rates on chemical potentials is a more general approach because it applies to far from equilibrium chemical reactions as discussed in many resources such as Beard and Qian (Beard & Qian, 2008; Qian & Beard, 2005). However, because of the challenges of determining the true chemical potentials of reactants, most chemical network analyses are based on the more familiar, and simpler, amount-driven reaction modeling.

OPB:Amount-Driven Flow Dependency

The most commonly modeled constitutive dependencies are based on the "law of mass action" whereby flow rates are assumed to be proportional to the difference in the amounts of flowing stuff in the sources and sinks of the flow. For example, chemical kinetic reaction rates can be modeled depending on the chemical concentrations of reaction participants according to the "law of mass action" kinetics for chemical reactions.

> OPB:*Amount-driven flow dependency*: A mono-constitutive dependency in which a flow rate between its players depends on the differences in the amount properties of its players.

Enzyme-mediated reaction flow. One of the more challenging resistive (flow) dependencies comprises the enzyme kinetic laws such as the Michaelis–Menten equations that were described in the previous chapter.

OPB:Capacitive Dependency

Capacitive dependencies (Figure 6.15) represent how, for example, the voltage (a force) across an electrical capacitor depends on the amount of electrical charge that it contains. Likewise, the air pressure in a balloon depends on the amount of air in the balloon.

• Capacitive force dependency
  • Fluid capacitive dependency
  • Electrical capacitive dependency
  • Mechanical capacitive dependency
  • Chemical capacitive dependency
  • Diffusional capacitive dependency
  • Electrochemical capacitance dependency

FIGURE 6.15   OPB:*Capacitive force dependency*: A constitutive dependency in which the force property of a participant depends on its amount.

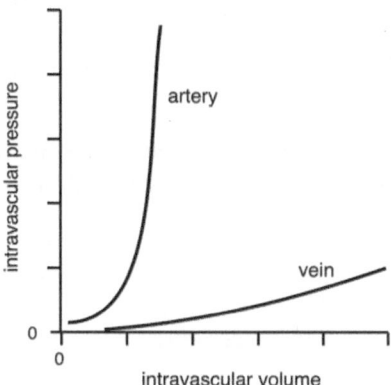

FIGURE 6.16   Distinctly nonlinear dependence of intravascular pressure on intravascular volume of arteries and veins.

Capacitive dependencies of force and amount are analogs of Hooke's law that, as first formulated, described a linear dependence of force on displacement of an elastic element such as a coil spring. However, except for the linearization of small displacements, biomechanical tissues and structures typically exhibit distinctly nonlinear force–displacement dependencies. For example, the pressure–volume relations of arteries and veins (Figure 6.16) that compare the volumetric stiffness of thick-walled arteries to thinner-walled veins.

OPB:Inductive Dependency

Inductive dependencies represent relationships between the momentum of an entity and its velocity.

OPB:*Fluid inductive dependency* (Figure 6.17) occurrences can represent the acceleration and deceleration of pulsatile blood flow through the aortic valve. OPB:*Mechanical inductive dependency* can represent occurrences of rapidly accelerating muscle and bones during musculoskeletal motions.

• Inductive dependency
  • Mechanical inductive dependency
    • Lineal mechanical inductive dependency
    • Rotational mechanical inductive dependency
  • Fluid inductive dependency
  • Electrical inductive dependency

FIGURE 6.17 An OPB:*Inductive dependency* is a constitutive dependency between an entity's momentum and rate properties.

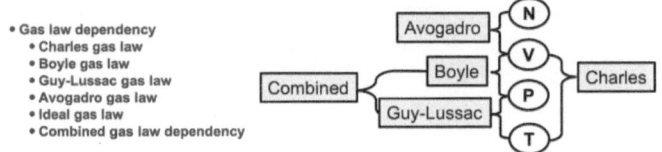

• Gas law dependency
  • Charles gas law
  • Boyle gas law
  • Guy-Lussac gas law
  • Avogadro gas law
  • Ideal gas law
  • Combined gas law dependency

FIGURE 6.18 (Left) OPB:Gas law dependency classes that represent five ideal gas laws that apply to properties of a portion of an ideal gas where

N = number of molecules in a portion of gas OPB:*Molar amount*
V = volume of space occupied by the gas OPB:*Fluid volume*
P = pressure of gas OPB:*Absolute pressure*
T = temperature of gas OPB:*Temperature*.

OPB:Gas Law Dependency

The classic gas laws were first described in the late 19th century by various scientists who noted proportionalities between physical property measures of a single portion of an *ideal gas*, i.e., a single participant. Each gas law bears the name of its discoverer and represents the proportionality among the various physical measures as detailed in Figure 6.18. OPB represents six gas laws, each as a subclass of OPB:*Gas law dependency*: (1) OPB:*Avogadro gas law*, (2) OPB:*Boyle gas law*, (3) OPB:*Charles gas law*, (4) OPB:*Guy-Lussac gas law*, (5) OPB:*Ideal gas law*, and (6) OPB:*Combined gas law*. Each gas law subclass declares the appropriate "property player" in the dependency as shown in Figure 6.18.

> OPB:*Gas law dependency*: A constitutive dependency of dynamical properties of a portion of an hypothetical ideal gas at thermodynamic equilibrium.

The aforementioned constraints and constitutive dependency classes apply to continuants and processes within a single domain and at a single spatiotemporal scale. Of course, a fundamental mandate of the OPB is to represent multiscale, multidomain biophysical phenomena and so OPB implements dependency subclasses for each, as follows. Note that the "Combined" gas law combines the laws of Boyle and Guy-Lussac.

OPB:*Dual Constitutive Dependency* between Two Continuant Participants

The essence of *multidomain* systems and their processes is that the physical properties of entities in one domain depend upon the physical properties of entities in a different domain. The contractile *force* of a muscle cell depends on the chemical *concentration* of calcium ions in the cell's cytoplasm. The *firing rate* of a carotid artery baroreceptor depends on the *fluid pressure* of blood in the carotid artery. Here, we will introduce only three subclasses of OPB:*Dual constitutive dependency* as examples.

OPB:*Transducer Dependency*

The essence of *multidomain* systems and their processes is that the physical properties of entities in one domain depend upon physical properties of entities in a different domain. The contractile *force* of a muscle cell depends on the chemical *concentration* of calcium ions in the cell's cytoplasm. The *firing rate* of a carotid artery baroreceptor depends on the *fluid pressure* of blood in the carotid artery. Thus, a transducer is a device that converts measures or displacements in one domain into that of another (Figure 6.19). For example, a thermometer converts temperature into the height of a mercury column, the position of a dial thermometer, or the voltage in an electrical circuit. A sensory neuron transduces a fluid pressure or a muscle extension into its firing rate.

OPB:*Transporter Dependency*

A constitutive dependency of a flow rate property of one dynamical entity depends on the force property of another dynamical entity. For example,

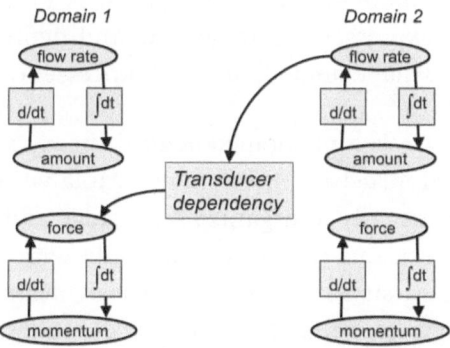

FIGURE 6.19 An OPB:*Transducer dependency* is a constitutive dependency that represents the dependence of, for example, the value of a dynamical property in one domain on a dynamical property in a different domain.

the flow rate of the membrane sodium-potassium exchange pump (NKP; NaKATPase) depends on the chemical potential (a force) of the conversion of ATP to ADP to provide the energy for the cotransport of sodium and potassium ions across cell membranes.

OPB:Transformer Dependency

A transformer constitutive dependency represents the flow rates of two role players in a single dynamical domain that are coupled to the respective forces of the role players. The familiar example from electrical engineering consists of two electromagnetic coils with enmeshed fields such that the AC voltage across one coil is transformed up or down depending on the number of "turns" in each coil. This occurs such that the force-times-flow (voltage-times-current, i.e., power) in one coil is scaled to the force-times-flow in the other. A playground teeter-totter is a mechanical transformer which has analogs in musculoskeletal levers such as the biceps muscle pulling on the forearm.

The OPB's representation of the myriad constitutive dependencies is nowhere near complete, even if one should imagine, and seek to represent, such a complete set. However, now we turn from representing mechanical details of physiological processes to representing the broader constraints of thermodynamics.

OPB:THERMODYNAMIC ENTITY

The development of the OPB has been motivated and inspired by the early work in network thermodynamics resulting in formulations such as the "tetrahedron of state" and the computational methods of energy-bond graph analysis. Thermodynamics is recognized as a foundational physical science because it expresses empirical laws and defines constraints that govern how and to what extent all physical processes occur.

> Thermodynamics is a phenomenological theory and, as such, is a purely formal structure. It offers no "explanation" of physical events, but serves only to organize knowledge and establish relationships between quantities…it deals only with the initial and final equilibrium states of a system; it provides no information about the dynamical behavior between these states.
>
> *(Oster et al., 1971)*

A foundational aspect of the physical realism of OPB is the recognition and representation of thermodynamic energy and entropy as continuants because they are key measures of and constraints on how continuants change during processes. For example, chemical reactions only occur with the expenditure of chemical potential energy and the corresponding increases in entropy. Hence, the OPB includes classes representing key thermodynamic quantities such as OPB:*Portion of energy* and OPB:*Portion of entropy* as OPB:*Physics continuants.*

As a pragmatic ontology of biophysical system dynamics, it is far beyond the scope of the OPB to represent the many profound formulations and implications of thermodynamics. Rather, we have focused on those definitions and laws that express dependencies, constraints (e.g., conservation of energy), and definitions as needed to represent classical thermodynamic laws and definitions to (1) annotate and reason over energy-bond graph models and (2) to propose and define OPB:*Physical process* as the flow of thermodynamic energy.

Thermodynamic energy is one of the four conserved quantities (energy, mass, charge, and momentum) of classical physics; yet, quantities of energy cannot be directly measured, so, and are known only by computations on other dynamical property values. Nobel Laureate Richard Feynman describes an energy conservation law as

> ...a conservation law means that there is a number which you can calculate at one moment, then as nature undergoes its multitude of changes, if you calculate this quantity at a later time it will be the same as it was before, the number does not change.
>
> *(Feynman, 1994)*

This is all the more remarkable because of the many forms and manifestations of thermodynamic energy and the universal law of the conservation of energy whereby amounts of energy can vary independently within a closed system; yet, their sum remains constant.

Thermodynamic Properties and Dependencies

Of particular interest and usefulness for system dynamical analysis are the subclasses of OPB:*Thermodynamical entity* that include classes for OPB:*Portion of energy*, OPB:*Portion of entropy*, and OPB:*Thermodynamic rate property* as listed in Figure 6.20.

The schematic diagram (Figure 6.21) represents the schema by which three thermodynamic properties (amounts of kinetic energy, potential energy, and heat) are defined in terms of the values of the four dynamical properties (force, flow rate, amount, and momentum):

- **Thermodynamical entity**
  - **Portion of energy**
    - **Portion of potential energy**
    - **Portion of kinetic energy**
    - **Portion of total energy**
  - **Portion of entropy**

- **Thermodynamic rate property**
  - **Energy flow rate**
    - **Heat flow rate**
    - **Total energy flow rate**
    - **Kinetic energy flow rate**
    - **Potential energy flow rate**
  - **Entropy flow rate**

FIGURE 6.20 (Top) OPB:*Thermodynamic entity* instances represent various portions of energy and entropy. (Bottom) OPB:*Thermodynamic rate property* instances represent rates of flow of energy and entropy.

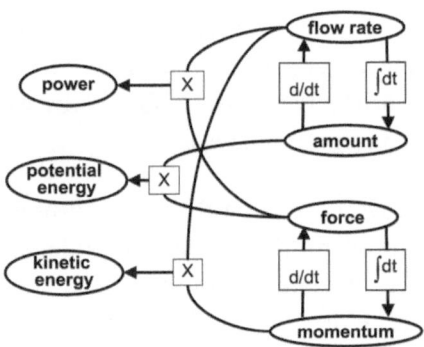

FIGURE 6.21 Schematic representation *OPB:Thermodynamic properties* (left) as they relate to the four OPB:*Dynamical property* classes (right).

1. An amount of kinetic energy (OPB:*Kinetic energy*) is the temporal integral of the product of an entity's momentum times its flow rate; e.g., the momentum of a mass times its velocity.

2. An amount of potential energy (OPB:*Potential energy*) is the product of an entity's amount property times its force property; e.g., the force on a spring times its displacement.

3. An amount of heat energy (OPB:Heat energy) is the temporal integral of a flow rate and force property; e.g., due to frictional drag forces on a moving surface.

Thermodynamics Is Universal

The appeal of thermodynamic-based computational methods is that they offer a universal bedrock theory that spans all dynamical scales and domains. The downside is that there are such a plethora of formulae and rules that are used to express the total energies of dynamical entities and the net energy flow rates of dynamical processes. To maintain the scope of the OPB, we have chosen to limit its representation to those properties and dependencies that are essential for representing network models based on thermodynamics of chemical kinetic (Oster et al., 1973; Perelson, 1975; Qian & Beard, 2005), mechatronic (Mikulecky, 1983), and multiscale biophysical systems which are based on the principles of conservation of energy.

The basic thermodynamic properties and their definitions and dependencies in the OPB constitute a very select subset of thermodynamic entities, dependencies, and relations that are used for analyzing dynamical systems. Anyone who has delved into and studied thermodynamics has found a science that is powerful in its generality and applications to all manner of phenomena based on multiple dynamical, temporal, and constitutive perspectives.

OPB:PHYSICS PROCESS

The Basic Formal Ontology (BFO) established early on the basic ontological framework for representing a process: something that occurs during a discrete temporal interval that begins at a start time, terminates at an end time, and during which participants in the process undergo changes (Smith, 2013). More precisely, the BFO defines *Process* as "an occurrent entity that exists in time by occurring or happening, has temporal parts, and always depends on some (at least one) material entity" (as a participant) (Arp et al., 2015). This fundamental idea underlies many biomedical ontologies including the Gene Ontology "biological process" branch (see Chapter 2). Below, we describe two other BFO-based process ontologies that represent, respectively, industrial manufacturing processes and biological cell processes.

First, the Process Specification Language (PSL) and its ontological approach define physical processes as discrete events in phenomenological manner by which continuant instances associate, dissociate, or change ontological type (Ozgovde & Gruninger, 2010). The PSL was developed to represent manufacturing processes as use cases include, for example, two manufactured parts that are joined to form a part of a different type, or, a part is painted or heat-treated. PSL defines a process as a "repeatable behavior whose occurrences cause continuants to undergo changes" and treats processes as discrete events that occur during a span of time.

Second, the Cell Behavior Ontology (CBO; Sluka et al., 2014)) adopts an OBO-based ontological view similar to PSL in that discrete continuants participate in processes by which they may be created/destroyed, attach/detach, and move across boundaries between anatomical compartments. CBO defines an array of CBO:*Fundamental physical processes*, such as CBO:*Movement* and CBO:*Molecular process*, that can be used to annotate observation and model behaviors. CBO implements a set of object–object relations such as *precedes* and *preceded_by* for representing the temporal ordering of processes and *participates_in* and *has_participant* for linking processes to the physical continuants that are their participants. CBO also defines an array of physical properties (CBO:*Physical object quality*) including subclasses of CBO:*MechanicalProperty* and of CBO:*ChemicalProperty*.

Whereas BFO, PSL, CBO, and others aim to represent and describe processes of complex systems, they offer only scant representation of the (classical) physics that represents and explains how such processes occur. Thus, the main aim for establishing the OPB:*Physical process* class is to recognize and represent the physics-based properties and the physical laws and dependencies that are used to explain and compute with biomedical hypotheses.

"Process" – Ontological Scope, Participation, and Properties

Our interest in physiology and biophysics stems from the fact that things happen, things change, and things can compose and recompose with other things. These occurrences are defined and classified in upper ontologies such as BFO as *occurrents* in which physical things (*continuants*) participate and change during the process. The BFO definitions are broad enough to encompass our definition of OPB:*Physics process* as "A physics occurrent that is the flow or exchange of matter, charge, or energy among dynamical continuants that are participants in the process." This definition applies to continuant participants that exist for the duration of the process, do not change type or class, and can be analyzed in a system dynamical framework as described earlier (Chapter 5).

OPB:*Physics Process* – Observable Process Events

OPB defines additional subclasses of OPB:*Physical occurrent* to represent the happenings that occur during physical processes (i.e., changes) as subclasses of OPB:*Physics process* that distinguishes discrete events (OPB:*Physics event*) that occur in a temporal instant that can represent, for example, the value of a physical property, as measured or computed, for some moment in time (Figure 6.22).

⋅ Key to any dynamical analysis is measuring and recording the time course of occurrences that are represented as instances of OPB:*Physics trajectory* defined as "A *physics change* that is the time course of changes, either discrete or continuous, in a physics process or in the physical state of a process participant." The key subclass OPB:*Physics change* is described below. Also critical for biophysical analysis are classes that represent physical happenings such as events (OPB:*Physics event*) that demarcate the start of a process, or the minimization or maximization of some property value (e.g., weight, velocity) during a process (OPB:*Property value trajectory*) as might be displayed on an oscilloscope or graph. This identification of dynamical process with the flow, expenditure, and dissipation of thermodynamic energy is motivated by the empirical observation that no processes occur – i.e., nothing happens – absent the flow of energy and/or information among process participants. This is true across spatiotemporal scales from chemical reactions and biophysical networks as well as for the self-organization that occurs during organismal development and evolution that occur via non-equilibrium energy dissipation in complex systems (Kauffman, 1993).

Changes in the value of one or more system properties over a span of time is evidence that a physical process is occurring. However, such changes are

- Physics process
    - Information process
    - Physics event
        - Continuant state event
            - Entity existence event
            - Structural change event
        - Property value event
            - Threshold value event
            - Extreme value event
        - Process occurrence event
            - Start process event
            - End process event
    - Physics trajectory
        - Event trajectory
        - Property value trajectory
        - Continuant state trajectory
    - Dynamical process

FIGURE 6.22 The class hierarchy of OPB:*Physics process* that is a subclass of OPB:*Physics occurrent*. Subclasses of OPB:*Dynamical process* are expanded.

- Dynamical process
    - Capacitive force process
        - Chemical capacitive process
        - Fluid capacitive process
        - Mechanical capacitive process
        - Electrical capacitive process
        - Diffusive capacitive process
    - Inductive process
        - Fluid inductive process
        - Mechanical inductive process
        - Electrical inductive process
    - Constitutive flow process
        - Amount-driven flow process
        - Potential-driven flow process
        - Transport flow process
    - Transformer process
        - Diffusive transformer
        - Mechanical transformer process
    - Transducer process
        - Chemo-diffusional transducer process
        - Chemo-mechanical transducer process
        - Fluid-mechanical transducer process

FIGURE 6.23   Part of the OPB:*Dynamical process* class hierarchy.

not the process itself. Rather, the OPB defines OPB:*Physics process* as "an occurrent that is the flow or exchange of matter, energy, and/or information among dynamical entities that are participants in the process." This definition reflects the thermodynamic foundations (OPB:*Thermodynamic entity*) of physical processes and is bound by the physical rules and computations that are occurrences of OPB:*Physical dependency.*

This definition of the physical process is based on the supposition that nothing happens in the physical world absent the flow and dissipation of energy and that causal relations among process participants can be interpreted as exchanges of energy and/or information. Therefore, the definitions and classifications of OPB:*Dynamical process* classes (Figure 6.23) map directly to subclasses of OPB:*Physical dependency* as described above with the recognition, in each case, that each such dependency describes a relation between a rate and state property whose products is an amount or flow rate of energy.

The OPB:*Dynamical process* classes represent a broad, but hardly complete, range of important, basic classes that is, however, incomplete when one confronts the challenges of the diversity and intricacy of biological processes.

OPB – STATE OF DEVELOPMENT AND FUTURE

The OPB is, by no means, complete. Our long-term goal has been to develop OPB as a comprehensive ontological representation of system dynamics that can usefully represent the physical things and processes that comprise biophysical systems. We have sought to develop a tool that suffices for several useful applications including the annotation and mapping of biophysical data and modeling resources, functional qualitative reasoning about biophysical models, and the encoding of biophysical system models for quantitative computation.

We have been challenged by the novel and evolving technology of biomedical ontology to represent first, the tenets and theories of engineering system dynamics in a concise and accurate fashion, and to face the challenges of the subtle complexities and essential non-linearities of biophysical systems.

CHAPTER 7

OPB-Based Semantic Modeling

A PRIMARY AIM IN DEVELOPING the Ontology of Physics for Biology (OPB) was to create a logical, computable framework for organizing mechanistic biological knowledge across physical scales and biological research domains. This framework has provided the foundation for the development of novel applications and techniques that leverage such knowledge, including the annotation and composition of biosimulation models, analysis of mechanistic biological networks using automated reasoning, and the semantic integration of biosimulation models and data. In this chapter, we discuss several applications of the OPB that illustrate its utility and that highlight the use-cases that have driven its development.

ANNOTATION OF BIOSIMULATION MODELS: CURRENT PRACTICES

Researchers in the field of biological modeling and simulation use a wide variety of computational formats and software platforms to build and test their models. Consequently, models developed by one group of researchers often do not interoperate with those developed by another. Furthermore, modelers do not use a standard set of identifiers to indicate the biological meaning of data structures in their source code. Thus, one modeler might use the identifier "x" to indicate heart rate in one model; another modeler might use it to indicate extracellular sodium concentration. The absence of such naming conventions makes it more difficult to repurpose

DOI: 10.1201/9780429469961-7

biosimulation models across research groups, a task that is becoming increasingly difficult as computational models become more complex.

One way to address this challenge is to annotate models so that the meaning of their contents is exposed in a machine-readable way. If such annotations could be standardized across modeling formats and software were built for reading/writing the annotations, then modelers aiming to repurpose models developed by others could easily access the biological meaning of a model's content and more quickly align the externally-developed model with their own. They would be able to readily assess where the two models overlap in terms of the biology represented, allowing them to focus on critical differences between the models that must be resolved before merging them together.

We refer to annotations that capture the biological meaning of a model's content as semantic annotations. In addition to their utility in the domain of model alignment and composition, semantic annotations are also a key ingredient for developing advanced search and retrieval tools within the biosimulation domain. Many biosimulation models are now publicly available in curated repositories. For example, the BioModels repository (Malik-Sheriff et al., 2020), maintained by the European Bioinformatics Institute, contains over a thousand models encoded in the Systems Biology Markup Language (SBML). The Physiome Model Repository (Yu et al., 2011), maintained by the Auckland Bioengineering Institute at the University of Auckland, contains hundreds of models encoded primarily in CellML. Given the differences in the modeling formats supported between these repositories, and the fact that the two repositories do not share a standard approach for representing model metadata, it is difficult for researchers to search over these repositories simultaneously. Semantic annotations offer a solution: if models in both repositories are annotated in a consistent, machine-readable fashion, then that metadata can be collected from both repositories, integrated, and made searchable by model discovery tools. Given the multitude of modeling formats used by modelers, semantic annotations provide a valuable common ground for capturing the biological meaning of what models simulate.

To make this common ground a reality, we have collaborated with community members of COMBINE (COmputational Modeling in BIology NEtwork: https://co.mbine.org) to develop a standardized approach for linking annotations to biosimulation models, independent of the modeling format used. These efforts culminated in the publication of a community-ratified specification that provides detailed recommendations regarding the storage and annotation of models (Gennari et al., 2021). We discuss this effort in more detail at the end of the chapter in the "Standardized Annotations for Modeling Projects" section. In the following sections, we discuss model annotations in more detail, focusing on the methodology and tools used for precisely capturing the biological meaning of model components.

COMPOSITE ANNOTATIONS

Given the value of semantic annotations, our research group has focused on developing a comprehensive annotation strategy applicable to models across biological research domains and physical scales. We built upon established annotation strategies that link model elements to controlled terms from curated knowledge resources such as ChEBI, the Foundational Model of Anatomy (FMA), UniProt, and the Gene Ontology. For example, curators of the BioModels repository have used this annotation strategy – codified as the Minimal Information Required In the Annotation of Models (MIRIAM; Le Novere et al., 2005) – to define the contents of SBML models.

This strategy of linking a model element to a single knowledge resource term may be sufficient for efforts that focus on models within a narrow biological domain such as biochemical kinetics; however, the existing set of terms from publicly available, curated knowledge resources does not provide sufficient coverage for the concepts simulated across biosimulation models. In many cases, no knowledge resource term exists that precisely defines what is being simulated in a biosimulation model. For example, to our knowledge, no resource provides a term that represents cytosolic calcium ion concentration in a pancreatic beta cell, yet this concept is often represented in models developed as part of diabetes research.

Given this limitation, our solution is to use *composite annotations*: logical statements composed of several reference terms linked together to compose a precise definition for a modeled element (Gennari et al., 2011). The basic structure of a composite annotation consists of an OPB term that defines (1) the physical property represented by the concept and (2) the bearer of the property. To create a composite annotation for cytosolic calcium ion concentration in a pancreatic beta cell, we first examine the physical property that is represented. In this case, it is chemical concentration; therefore, we use the OPB term "Chemical concentration" as the first component of the annotation. Next, we determine the bearer of the property. In this case, it is the set of calcium ions in the cytosol of a pancreatic beta cell. Because no single knowledge resource term represents this exact physical entity, we must further decompose the concept into more fundamental terms: calcium ion, cytosol, and pancreatic beta cell. Fortunately, knowledge resource terms exist that represent these more fundamental concepts, and so we create a logically defined "composite physical entity" using the terms ChEBI:*calcium(2+)*, FMA:*Portion of cytosol*, and FMA:*B cell of pancreatic islet*. We use structural relations between the entities to indicate their relationship with each other. Figure 7.1 illustrates the full composite annotation.

It may be confusing as to why the FMA names the cytosol "Portion of cytosol". This is because the FMA's strict formalism is based on structural features, and objects without inherent 3D shape are "Portions". So, for example, in addition to FMA:*Portion of cytosol*, there is FMA:*Portion of blood* and FMA:*Portion of cerebrospinal fluid*. The FMA is primarily used to identify physical objects that might be sitting in front of you on a metal slab, so if you want to identify, say, the blood that is in the left ventricle, it is more ideologically correct to say that it is a "Portion of blood" rather than "Blood" in general; the latter would indicate a reference to all blood, everywhere.

The example composite annotation above represents a physical property of a physical *entity*, and many concepts represented in biological models fall into this category. This includes, for example, concentrations

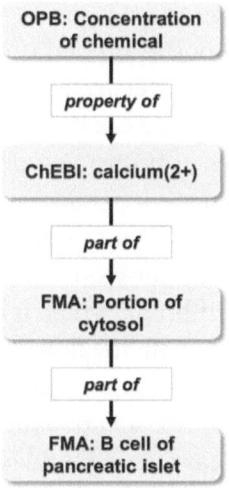

FIGURE 7.1 Schematic of a composite annotation for a property of a physical entity: the concentration of cytosolic calcium ions in a pancreatic B cell.

of chemicals, volumes of fluids, and density of solids. However, not everything simulated in a biological model is a property of a physical entity. For example, the rate of a chemical reaction is a property of a physical *process*. Consider a model that simulates the rate of the glucokinase reaction in glycolysis. How should this concept be represented as a composite annotation? Again, we decompose the concept into the property represented and the property bearer. In this example, the property is chemical flux – i.e., the amount of chemical produced or consumed per unit time. If this rate is quantified in a model with units of moles per second, then we use the OPB term "Chemical molar flow rate" for the physical property in the composite annotation.

Next, we consider the property bearer, which is the physical process that is occurring. It would be convenient if there was a knowledge resource that represented all physical processes that could occur, but none exists, and creating such a resource is unrealistic. Therefore, we employ a solution where we define the process by the physical entities that participate in it. In doing so, we specify which physical entities are energetic sources and sinks in the process to indicate its directionality. We also specify any participating mediators: physical entities whose properties influence the magnitude of the process rate but whose amounts remain unaltered by the process. For example, glucokinase (also known as hexokinase-4) is the enzyme that mediates an important reaction in the glycolysis pathway. While it is a critical participant in the reaction, its stoichiometry is zero. The reaction consumes the sources glucose and ATP and produces glucose 6-phosphate and ADP. Thus, the full composite annotation for this example can be represented using the relationships illustrated in Figure 7.2.

In addition to properties of physical entities and processes, the OPB defines two additional types of properties: properties of energy differentials and properties of physical dependencies. These properties are also found in biosimulation models. An example of a property of an energy differential is the fluid pressure of blood in a vessel. Another example is the voltage across a cell's membrane, which is often represented in electrophysiological models that simulate the flow of ions into and out of a cell.

To create a composite annotation for, say, voltage across the membrane of a pancreatic beta cell, we first identify the corresponding OPB physical property term. In this case, it is OPB:*Voltage*. Next, we identify the property bearer. As with physical processes, there is no comprehensive ontology that represents all energy differentials; therefore, we instantiate an unnamed differential that is defined by the physical entities that act as

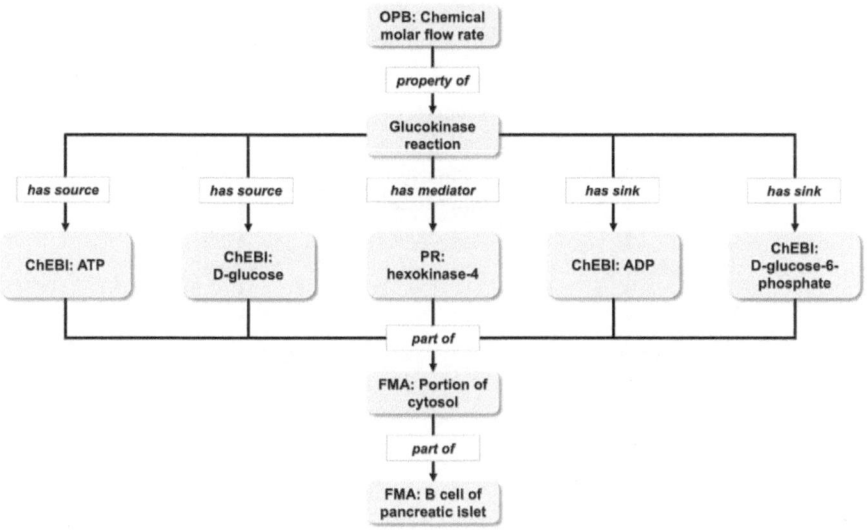

FIGURE 7.2 Schematic of a composite annotation for a property of a process: the chemical flow rate of a reaction that converts ATP and D-glucose into ADP and D-glucose-6-phosphate.

the differential's energetic sources and sinks. The voltage across the membrane of a cell exists due to a difference in ionic charge between the cell's cytosol and its extracellular milieu. Therefore, the physical entity acting as the differential's thermodynamic source is the set of charged ions in the cytosol, and the entity acting as the sink is the set of charged ions in the extracellular matrix. Figure 7.3 illustrates the full composite annotation that defines this concept.

Electrophysiology models simulate ion currents and membrane voltages that are due to the intracellular and extracellular concentrations of a specific type of ion. These so-called Nernst potentials (OPB:*Nernst reversal potential*) are defined as the membrane voltage difference given the intracellular and extracellular concentration of a specified ion, such as K⁺. To represent the Nernst potential due to potassium ions across a cell membrane, we would replace the more general term ChEBI:*ion* with the more specific ChEBI:*potassium(1+)* in Figure 7.3.

The last type of physical properties we discuss in the context of annotating biosimulation models are properties of physical *dependencies*. These properties represent the characteristics of mathematical or logical relationships between two or more physical properties. Thus, quantifying properties of dependencies requires the quantification of two or more

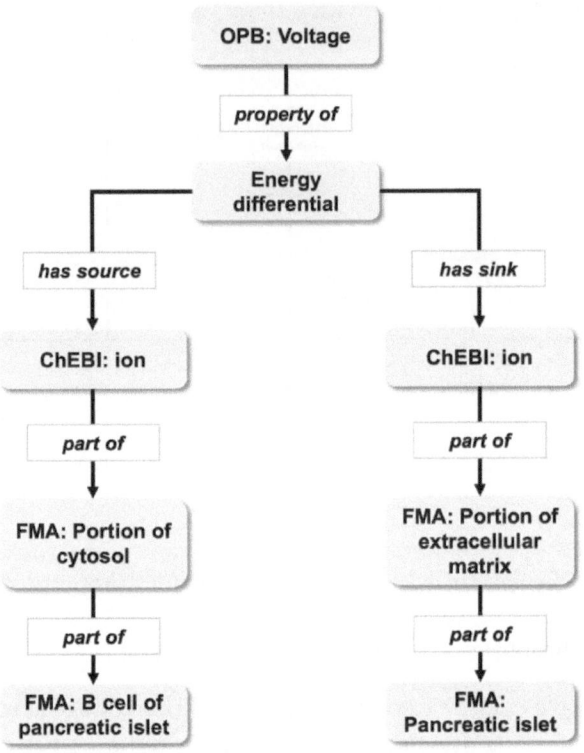

FIGURE 7.3 Schematic of a composite annotation for a property of an energy differential: the voltage resulting from the difference in ionic energy between the interior and exterior of a pancreatic B cell.

other physical properties. For example, OPB:*Fluid flow resistance* is the relationship between an OPB:*Fluid pressure* and an OPB:*Fluid flow rate*. This is the fluid analog of electrical resistance, which is the relationship of an OPB:*Electrical potential* to an OPB:*Charge flow rate*, and the purely linear version of this relationship is Ohm's Law. Many properties of dependencies associated with established physical laws have been named, such as electrical resistance or the Michaelis–Menten constant, but many formulas used in biosimulation models are created *ad hoc* and use curve-shaping constants with no canonical definition. For example, the generic weighting factors used to fit a polynomial to a non-linear pressure–flow relationship do not have biological meaning apart from their use in the dependency relating pressure and flow. As we will discuss in the next section, annotating such factors in biosimulation models is more difficult and, for many modeling use-cases, unnecessary.

From these composite annotation examples, a comprehensive strategy for annotating biosimulation models requires a thorough, well-defined set of physical properties for use in composite annotations. Historically, this has been one of the primary motivations for developing the OPB, which now contains a broad collection of physical property terms for use in annotating a wide variety of biosimulation models (see Chapter 6, "OPB:Physics Property – Observable and Computable" section). In the following sections, we discuss the utility of composite annotations in the field of biosimulation as well as the tools we have built that support their creation and application.

THE SEMSIM ARCHITECTURE

The primary driving use case for OPB development has been for thoroughly annotating the semantics of biosimulation models such that software tools can recognize where models overlap in terms of their simulated biological content, to make that content searchable, and to formalize that content such that automated inference tools can perform reasoning tasks over it. To achieve the latter two goals, we developed a logical framework for capturing the biological and computational content of biosimulation models in a format that is readily queryable and to which automated reasoners can be applied. We call this framework the Semantic Simulation (SemSim) architecture, and we use the Web Ontology Language (OWL) as the format for storing the contents of models within this framework. OWL is based on the Resource Description Framework (RDF), which is readily queryable using SPARQL, and OWL files can be reasoned over, using automated tools such as HermiT (Glimm et al., 2014). Using OWL also allows us to integrate knowledge from other OWL-based ontologies, such as the OPB and FMA to perform automated reasoning analyses over biosimulation models. For example, we have developed an approach for performing qualitative perturbation experiments over biosimulation models that uses knowledge from the OPB and automated reasoning tools (Neal et al., 2016). We discuss this application later in the chapter in the section titled "Qualitative Inference Using the OPB and SemSim Models".

Organization of the SemSim Architecture

At the highest level of organization, SemSim models divide the contents of a biosimulation model into its computational and physical aspects (Figure 7.4). This is reflected by the top-level OWL classes in a SemSim model: *Computational_model_component* and *Physical_model_component*.

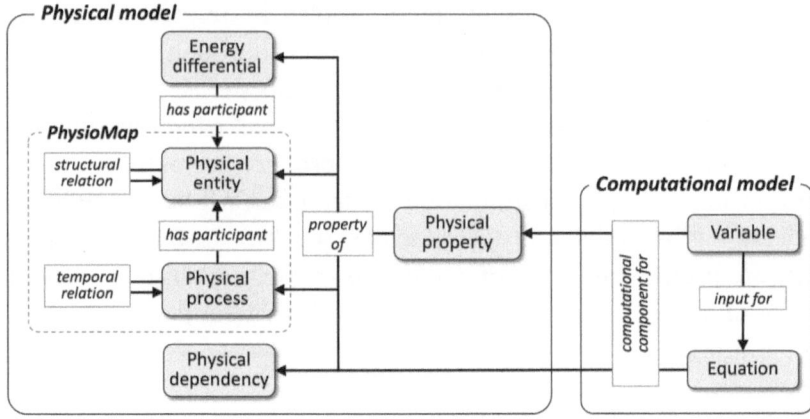

FIGURE 7.4 The SemSim architecture represents the biological elements of a simulation model that consists of variables and equations that represent the values of physical properties and the physics-based mathematical equations that represent how those values depend upon one another.

The former represents elements of a model that relate to the mathematical or logical representation of physical quantities or qualities. Elements in the latter class represent, in whole or in part, the physical meaning of a model's computational elements.

In other words, the *Physical_model_component* class captures the model's biological aspects, including the physical processes represented by the model, the physical entities that participate in them, and the physical properties simulated in the model. The subclasses of *Physical_model_component* are organized and interrelated according to the formal structure of the OPB. For example, *Physical_processes* have *Physical_entities* as participants, and instances of both can have *Physical_properties* that represent measurable quantities in a model. The relationships between these classes provide the structure for composite annotations, as discussed in the previous section. The *Computational_model_component* class captures the mathematical and logical relationships that determine how the quantities of these physical properties change over time. Instances within this class include model variables and the equations used to compute them when the simulation is executed.

The SemSim architecture was designed to represent the computational and biological aspects of simulation models across a wide variety of research domains and simulation formats. It can explicitly represent the biological processes of, for example, chemical reaction networks, electrophysiological phenomena, or tissue-level cardiovascular hemodynamics. This is accomplished by linking the *Computational_model_component* elements of a SemSim model to logical descriptions of what they represent, biologically, using composite annotations. The SemSim architecture has also been designed as a format that can represent the content of models encoded in a variety of model-interchange formats such as SBML and CellML. Because models in these formats can all be represented as SemSim models, the SemSim architecture can act as an intermediary for translating between these modeling formats.

The SemSim Application Programming Interface (API)

We have implemented the SemSim architecture as the SemSim Java API, which allows users to programmatically create, read, edit, and write SemSim models. This tool can aid researchers in such tasks as applying composite annotations to models, translating between modeling formats, and assessing the biological content of a model. The SemSim API's object model reflects the class hierarchy of SemSim OWL files. For example, the Java classes in the API are primarily grouped into either computational model components or physical model components. The physical model components include physical entities, processes, properties, and dependencies. The computational components include data structures such as decimals, unit definitions, and discrete event assignments.

This API is freely available at https://github.com/SemBioProcess/SemSim-Java-API, and a C/C++ version of the API is currently under development. It is our hope that these APIs will provide the broader biological modeling community with a valuable tool for managing and editing semantic annotations on models across modeling formats. We intend the API to reflect the recommendations of the broader community on how to serialize model annotations in a standardized way (Neal et al., 2019), including support for reading and writing annotation files within OMEX archives (Bergmann et al., 2014), an emerging standard for sharing models and associated data needed to reproduce simulation studies.

PHYSIOMAP ARCHITECTURE

The collections of composite annotations in a SemSim model can be analyzed to create informative network visualizations of the processes simulated in a model and the physical entities that participate in them. We call these visualizations PhysioMaps (Cook et al., 2013) and they can be derived from the process–entity relationships asserted in a SemSim model

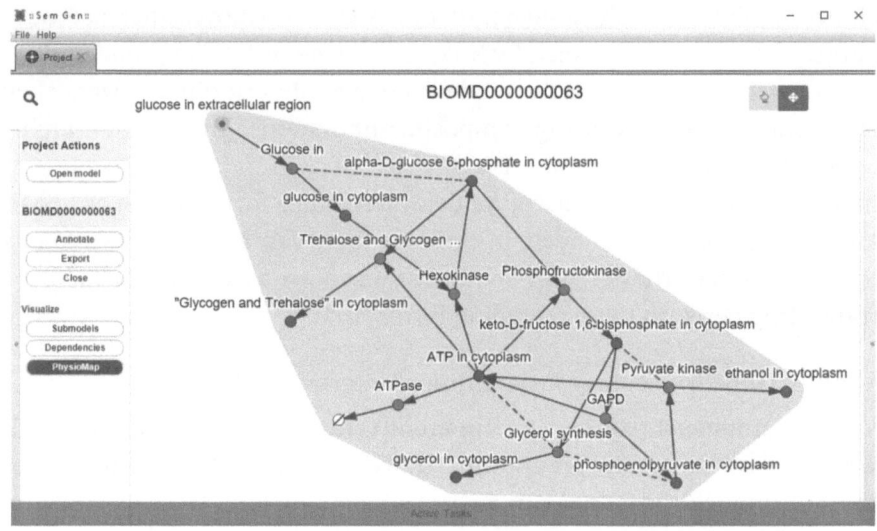

FIGURE 7.5 SemGen screenshot showing a PhysioMap of BioModels model BIOMD0000000063, which simulates the yeast fermentation pathway.

(Figure 7.4, dashed box). PhysioMaps generalize the network diagrams often used by systems biologists that show the structure of a chemical network. However, whereas chemical network visualizations show chemical species and the chemical reactions in which they participate, PhysioMaps are more general: they show physical entities and the physical processes in which they participate. PhysioMaps, therefore, are a parent class of chemical network visualizations. They can also show the network structure of, say, a hemodynamics model as well as a chemical network model.

Representing Processes in PhysioMaps

Once a SemSim model is created and annotated so that its physical process and entities are completely specified, PhysioMaps can be used to visualize the relationships between entities and processes in commonly-used node-and-edge visualizations. For example, the SemGen software tool (discussed next) uses d3 technology [https://d3js.org/] to display PhysioMaps of SemSim models using force-directed graphs (Figure 7.5).

SEMGEN

One of the primary motivations for developing the OPB and the SemSim architecture is to formally organize the physiological knowledge in bio-simulation models so that software can automatically recognize when models simulate the same (or similar) biological phenomena. Such capabilities give modelers more powerful methods for retrieving models of interest from repositories, discovering models relevant to their research, recognizing coupling points between models when composing them into larger systems, decomposing models into smaller systems, and for establishing links between models and empirical physiological data. Motivated by the goal of enhancing model composition and decomposition, we developed a software tool called SemGen, available on GitHub at https://github.com/SemBioProcess/SemGen (Neal et al., 2015, 2019; Sarwar et al., 2019). This tool allows users to annotate models with our composite annotation framework and leverages those annotations to enhance model composition and decomposition tasks. As part of the SemGen project, we have created a set of guidelines for annotating models with the tool, available at https://github.com/SemBioProcess/SemGen/wiki/Annotation-protocol.

SemGen provides capabilities for annotating SBML, CellML, MML (for JSim models), and SemSim models, as well as node-and-edge-based visualization tools for composing or decomposing models. SemGen also provides several visualization options for exploring a model's mathematical and biological content, including PhysioMaps (Figure 7.5). In previous work, we and others have described how SemGen can be used to accomplish complex model merging tasks in a semi-automated fashion that reduces the hand-coding required to build complex, composite systems (Beard et al., 2012; Neal et al., 2015; Shahidi et al., 2021).

The SemSim API undergirds SemGen, and when merging annotated models together, SemGen uses the functions of the API to recognize where two models overlap in their biological content. For example, if two models both contained variables annotated with identical composite annotations for cytosolic calcium concentration in a pancreatic beta cell, SemGen flags this overlap as a point of coupling between the two models. SemGen lists these overlaps so that the user can choose which computational representation of the biological property they would like to preserve in the merged system. For example, the value of calcium concentration in one model might be a static constant whereas in the other, it might depend on cytosolic reactions.

SemGen reveals these differences so users can avoid duplicating biological content in the merged model and can choose variable formulations that best harmonize with their modeling goals. SemGen and its underlying API thereby support "white-box" model composition, where all points of connection between merged models are exposed and the interface between models is determined at the time of coupling based on their overlapping biological features. This approach is in contrast to "black-box" composition wherein the models possess previously-defined interfaces that limit the potential coupling points between models.

We have previously discussed the characteristics of these approaches as well as their advantages and disadvantages (Neal et al., 2014). We have adopted a white-box approach that omits predefined interfaces because establishing such interfaces presumes knowledge of all the ways users might use the model in a composition task. Because this is not generally knowable, we have opted for an approach that provides a maximum level of flexibility for interfacing models into composite systems.

QUALITATIVE INFERENCE USING OPB AND SEMSIM MODELS

To understand and predict the behavior of biological systems, researchers often model their system of interest to organize their thinking about how the system's subcomponents interrelate. Such models may be fully quantitative if sufficient quantitative data is available for constraining the model. In the absence of full quantitative constraints, systems researchers can still construct qualitative models that represent a system's behavior in a categorical way. Such qualitative models might, for example, indicate whether the chemical species amounts in a signaling network increase or decrease given an increase in one of the participating species. While the model would not be able to quantify *how much* the species increase or decrease, it would nonetheless provide a valuable structure for representing a system before the system has been quantitatively characterized. Such models can be useful tools for representing and testing qualitative hypotheses about a system's behavior.

In Chapter 5, Section "Qualitative, Discrete Causal Methods", we described the Chalkboard software (Cook et al., 2007) that implements qualitative reasoning over cell physiological systems that are represented as node–arrow diagrams wherein nodes represent *continuants* (e.g., a kinase-site on an enzyme molecule) and arrows represent the *processes* in which the continuants participate (e.g., a phosphorylation reaction process). Each continuant (i.e., icon) has a *state property* such as its *amount* or *activation state* and each process (i.e., arrow) has a *rate* property that reflects its activity rate. Thus, Chalkboard can perform qualitative perturbation experiments that reveal how an increment (or decrement) of a given property value (e.g., a substrate concentration or reaction rate) propagates through the causal network to affect any other property value.

A Chalkboard user can, thus, interrogate as a system model using the "Path trace" tool to, one-by-one, perturb the state property of a selected node (continuant) or the rate property of a selected arrow (process). Results are graphically displayed as up- or down-arrows according to the propagated perturbations (increment or decrement) through the causal network. In an OWL-encoded SemSim model, such causal links are represented by instances of physical dependencies linked to instances of physical properties by OPB:*hasPropertyPlayer* relations. Thus, we have leveraged the OPB's formal theory of biological mechanistic phenomena to develop an automated approach for inferring the qualitative consequences of perturbations to SemSim models, thus providing a general solution for investigating qualitative perturbations on models that represent systems-level hypotheses.

Our approach is to (1) comprehensively annotate a biosimulation model in accordance with the SemSim annotation protocol (https://github.com/SemBioProcess/SemGen/wiki/Annotation-protocol), (2) store the model as a SemSim OWL file, and (3) use automated reasoning tools such as HermiT (Glimm et al., 2014) to identify how the physical properties simulated in the model, when increased or decreased in amount, qualitatively impact the physical properties that are dependent on them (Neal et al., 2016). For example, an increment of an enzymatic reaction rate would increment the concentration of each product, which would increment the rates of follow-on reactions that use the products as reactants. This is the sort of qualitative reasoning investigators and educators use for understanding and reasoning about the function of systems that is valuable for proposing and testing theories of physiological and pathophysiological function. Our qualitative causal reasoning tool can automatically infer these relationships based on the information contained in the model's OWL file and a set of SWRL rules (Horrocks et al., n.d.) that indicate conditions where a physical property has a positive qualitative influence on another, and where it has a negative influence. For example, one of the SWRL rules used to determine positive/negative role player status states that if a process rate (such as that of a chemical reaction) is solved using the property of a thermodynamic source of that process (such as a reactant in first-order kinetics), then the source property is a positive player in the process rate dependency.

When we run an automated reasoner on the OWL file representation of the model that includes these SWRL rules, we can establish how a qualitative perturbation in any given physical property in the model affects those properties that are immediately dependent on the property. However, these effects must be propagated throughout the remainder of the model's full dependency network to determine the global impact of a perturbation. To perform this propagation, we use a custom algorithm implemented in Java similar to colored petri nets (Jensen & Kristensen, 2009) that determines the qualitative effect of the initial perturbation on the properties that depend on the perturbed property, then we determine the subsequent downstream effects of those perturbations. We do this iteratively across the model's dependency network and determine which properties will be increased or decreased by the perturbation, which will be unaffected, and which will be affected in an ambiguous fashion. Loops within the dependency network may result in positive or negative feedback on a given physical property. In the case of positive feedback, the qualitative effect on the property would remain the same. Whereas traversing a negative feedback loop would introduce an effect on a property that is in the opposite direction of a previous upstream effect. Although this results in an ambiguous qualitative state for the property, the identification of such negative feedback loops can be valuable for understanding the network characteristics of a biological system.

In terms of the computational cost of this qualitative analysis approach, the bottleneck is the step where the automated reasoner classifies the SemSim OWL file. Models of median size within the BioModels database can be classified in a manner of seconds with a commercially available laptop but may take considerably longer for larger models. Despite this, the tool demonstrates how the OPB and the SemSim architecture can be leveraged to perform complex reasoning tasks on biosimulation models and that the OPB's scope extends its utility beyond domain-specific models: perturbation experiments can be performed with our qualitative reasoning tool for models of chemical networks, electrophysiological transport, tissue-level hemodynamics, and phenomena from many other research domains.

VISION FOR A SEMANTICALLY-INTEGRATED PHYSIOME

Given that the OPB represents high-level classes for organizing and reasoning over physiological phenomena at multiple scales, we envision using it as a centerpiece for integrating physiological knowledge across heterogeneous resources such as the quantitative and qualitative mechanistic models residing in various online repositories. We envision using the OPB to annotate these disparate resources, and then integrate the mechanistic physiological knowledge within a common knowledge base that allows users to explore physiological pathways across various biological levels of organization. Such a knowledge base for a given organism constitutes its physiome because, as discussed in Chapter 2, a physiome "describes the physiological dynamics of the normal intact organism and is built upon information and structure (genome, proteome, and morphome)" (Hunter & Borg, 2003).

The physiome concept has existed since the 1990s; however, the complete construction of any organism's physiome remains an unmet challenge. In our view, the fundamental problem is that there is no comprehensive, multiscale knowledge architecture for organizing and relating the entirety of an organism's physiology that accounts for the different representational formats used by physiological researchers to articulate their understanding of physiological processes. This is precisely one of the roles that the OPB is designed to play. It is not in the interest of single research labs to build a physiome for their organism of study, and few, if any, can do so, since accomplishing the task would require expertise in many subdisciplines of biological research. In our vision, a physiome could be built by aggregating and integrating the physiological knowledge that has been generated by various research groups and articulated in qualitative and quantitative models of physiological processes. Our approach would be to collect quantitative physiological models from public repositories such as BioModels and the PMR, annotate them in a consistent fashion with composite annotations, store the semantic and computational information from each model as a SemSim model, then integrate the axioms of these SemSim models into a unified knowledge resource. The composite annotations would act as the semantic glue that links physiologically synonymous components of different models together as they are added to the knowledge resource, allowing mechanistic pathways from different models to be coupled together to create larger super-pathways that potentially span scales.

A physiome constructed this way would provide a gene-to-organism perspective on an organism's physiology because its building blocks include models that span structural scales. Given sufficient coverage by extant mechanistic models, a user would be able to explore how glucose levels in the blood affect cardiac contraction and the flow of blood throughout the circulation. Because a semantically-integrated physiome would be built from SemSim models, and the SemSim format can represent or link to the computational aspects of the models they are derived from, a user would also be able to explore the different ways that a given physiological phenomenon of interest has been represented mathematically among the broader community of physiological modelers. Researchers could, for example, explore different formulations for a given enzymatic reaction of interest, or the different ways left ventricular contraction has been modeled. Thus, a semantically-integrated physiome would also allow users to discover models of interest across disparate model repositories, which is one of the long-standing goals of researchers developing biosimulation model- and data-sharing standards.

Constructing a semantically-integrated physiome would also give researchers the power to explore mechanistic connections between physiological features of interest. Users could query the physiome to discover the processes that relate, for example, a certain gene's activity to insulin secretion by pancreatic beta cells. Such queries might rely on shortest-path or minimal spanning tree algorithms to traverse the integrated process network within the knowledge base and discover mechanistic routes between physiological features of interest. Because these routes might traverse the content of multiple models, our resource could suggest model merging tasks that could be performed to generate new mathematical models that relate the user's physiological features of interest.

The lynchpin to the development of a semantically-integrated physiome is the annotation of the models that will comprise it. Currently, there are over a thousand annotated models within the curated branch of the BioModels repository, and the annotations therein can be readily converted into composite annotations used in SemSim models. However, there are well over 1,000 non-curated BioModels and models in other repositories, such as the PMR, that remain unannotated. Therefore, developing tools that accelerate model annotation and establishing quality control protocols that ensure consistent, accurate annotation across models are both important challenges for accelerating the creation of physiome-level knowledge resources.

Driven in part by these needs, members of the NIH-funded Center for Reproducible Biomedical Modeling project [https://reproduciblebiomodels.org/] are currently using SemGen to annotate CellML models in the PMR (Sarwar et al., 2019; Shahidi et al., 2021). This is being done with the goal of enhancing the visibility and discovery of published simulation models and to ensure that published models are readily comprehensible and reusable for new research projects. It is our hope that the number of consistently, thoroughly annotated models in public repositories increases substantially in the coming years, offering the opportunity to integrate the mechanistic physiological knowledge contained within them and realize one of the primary motivations for building the OPB: the construction of multiscale, multidomain knowledge bases of organismal physiology.

DATA ANNOTATION FOR REUSE

Another long-standing challenge in biomedical informatics is the development of a comprehensive approach for annotating the semantics of empirical data. This includes data generated by biological research labs and clinical data generated by health practitioners. Such data is annotated (if at all) using disparate schemes, not all of which interoperate. In our vision, modelers should be able to search data repositories for measurements they can use to parameterize and validate their models. Conversely, researchers and clinicians should be able to easily discover models that can be used to analyze the data they collect. One of the keys to achieving this is to annotate the data according to a common protocol.

We believe the corpus of composite annotations generated in the annotation of biosimulation models could provide a rich set of concepts to use for broad annotation of empirical data. For example, imagine such a resource called the Physiome KB where each unique composite annotation from a biosimulation model is assigned a unique URI. These identifiers could be used as machine-readable annotations on data items (much like a UniProt ID is used to indicate the meaning of a gene product in a transcriptomic data set). Thus, a researcher with publicly-accessible voltage measurements in a pancreatic beta cell could include a Physiome KB URI in their data set that points to a composite annotation for that concept. Similarly, clinicians could annotate physiological measurements using the same scheme.

Leveraging composite annotations for data annotation also has an advantage over schemes that rely on ontologically "flat" resources: Composite annotations are built using knowledge resource terms from ontologies where terms are linked to each other via class hierarchies, partonomies, and other topologies. We envision using these connections to create a robust, reliable metric for quantifying the semantic *similarity* between composite annotations. Thus, when a researcher searches a database annotated using the scheme articulated here, search engines can rank results based on similarity to the user's query and present data that not only matches the query exactly, but is also semantically related. For example, a modeler seeking to parameterize a hemodynamic model with a blood pressure measurement from the superior vena cava might also find pressure measurements from the systemic veins usable as a surrogate if vena cava measurements are not available. We envision leveraging the structural relationships in ontologies such as the FMA to recognize the anatomical relation between the superior vena cava and systemic veins, and then use that anatomical knowledge to recognize the similarity between blood pressure measurements across the corpus of searchable datasets. Establishing a semantic similarity metric for composite annotations is therefore a key challenge for realizing the benefits of a data annotation protocol based on composite annotations.

STANDARDIZED ANNOTATIONS FOR MODELING PROJECTS

As mentioned, one of the keys for developing large-scale physiological knowledge resources such as the Physiome KB is the annotation of a large corpus of biological models. However, even if such a corpus was available, the process of aggregating the semantic information in these annotations is hindered by the fact that different modeling formats use different schemes for storing annotations. We, along with other members of the COMBINE community, have therefore worked toward the development of a community-driven standard for representing annotations on models and other file types used in modeling projects. This work culminated in the ratification and publication of the Open Modeling and Exchange (OMEX) metadata specification, which provides technical guidelines for storing annotations on files in a readily-sharable modeling project (Gennari et al., 2021). OMEX-formatted archives, often referred to as "COMBINE archives", are basically zip files that bundle files associated with a modeling project. These archives can include model files, such as SBML or CellML-encoded models, as well as others that contain empirical data or visualization layout information.

The COMBINE community recommends using OMEX-formatted archives to share modeling project information among the broader biological modeling community, and the OMEX metadata specification that we have helped develop provides technical and policy guidelines for storing annotations on files within an OMEX archive. Building significantly from our work developing the SemSim framework and our composite annotation approach, the specification recommends that semantic annotations on archive contents are encoded as RDF-formatted triples stored in a separate file within the archive. The specification lays out the details for how to encode this metadata, which includes composite annotations. The specification recommends, as in the example composite annotations within this chapter, that OPB physical property terms are used when storing composite annotations in OMEX metadata files. This demonstrates an important contribution of the OPB to broader standardization efforts for model and data annotation and illustrates the practical utility of the OPB as a source of physical property terms that are relevant for the biosimulation modeling community.

As part of the Center for Reproducible Biomedical Modeling, we are actively working on developing software libraries that support the OMEX metadata specification (Welsh et al., 2021), including enhancements to SemGen and its underlying SemSim Java API. Our hope is that these tools will be useful to the modeling community for storing and sharing annotations on modeling projects, will help eliminate barriers that hinder the integration of semantic information across modeling formats, and will ultimately lead to large repositories of OMEX-formatted archives with standardized annotation information that can be readily aggregated into the kind of physiome-level knowledge resource we have described here.

CHAPTER 8

OPB Review and Possibilities

I N THIS FINAL CHAPTER, we both review the basic ideas of the Ontology of Physics for Biology (OPB) and emphasize its unique characteristics and potential for high impact in the realm of biological modeling. Our development of the OPB has been driven by the idea of representing physiological and biophysical systems in a formal, computable manner. Indeed, the original goal was to build a "Foundational Model of Physiology", as a companion to the Foundational Model of Anatomy. However, as our efforts progressed, we recognized the importance of capturing the fundamental physics of the dependencies among measurable physical properties of the entities that participate in physiological processes. Thus, our message that "It's all physics" and hence the name "Ontology of Physics for Biology".

Clinical practice and biomedical research are based on a common understanding that biological systems are physical systems that behave according to laws of physics. This understanding is the basis for scientific research, biomedical education, and clinical practice. However, the broad range of physical scales and the variety of physical domains present major challenges for learning, understanding, and treating biomedical problems and diseases. The scientific bases of these challenges are often introduced and taught by domain experts as a "bag of equations" – "$f = ma$", "$pV = nRT$", "$dV/dt = f(t)$", etc. – using a variety of physical perspectives, mathematical formalisms, and bioscientific nomenclatures. Since the 1950s, there is a proud history of math-based, system dynamical modeling of normal and

DOI: 10.1201/9780429469961-8

pathological biological systems that have contributed greatly to our understanding of health and disease.

Although accessible and embraced by students, researchers, and practitioners, this more traditional approach presents challenges to developing the intuitions and deeper understanding that are the necessary bases for representing and reasoning about multiscale, multidomain biophysical systems. Indeed, we recognized that the preponderance of system descriptions and inferences about biophysical and their behaviors occurs *verbally* in lectures, discussions, and publications without the formalisms required for physics-based mathematical models and analyses. The emergence of formal biomedical ontology in the mid-1990s suggested that a formalized semantics of classical physics could be the basis for representing, reasoning about, and computing on the structure and function of system dynamical models in physiology and biophysics. Although much work was done to create ontologies of biological entities (e.g., anatomy, as presented in Chapter 2), there has been a gap in ontologies that formally specify processes. In this text, we have introduced the OPB, to formally capture the ideas of processes as understood by bioengineers and biophysicists.

OPB USES AND APPLICATIONS

Ultimately, our goal in creating the OPB has been to develop a logical framework that will help integrate physiological knowledge across research domains and physical scales. Our primary use-cases have been the representation and integration of mathematical models of physiological processes. Researchers have used such models to articulate physiological knowledge for decades; however, these models have been encoded in a variety of languages that do not often interoperate, limiting access to the knowledge expressed within them as well as their reuse and integration with each other. Our aim has been to create a knowledge resource that would help bridge the various representations of quantitative physiological knowledge that have been, and continue to be, generated.

The OPB offers a comprehensive semantics for both qualitative and quantitative analyses of multiscale, multidomain systems and, as described in Chapter 7, has proved useful for annotating and reasoning about computational models in the context of the Physiome Project (Nickerson et al., 2016) and for the aims of the Center for Reproducible Biomedical Modeling. As its name suggests, the Center for Reproducible Biomedical Modeling is focused on promoting reproducible models, but key aspects of reproducibility are both understandability and interoperability. For these two aspects, models must be annotated against common semantics so that others can more easily understand and reuse the model. For many models, the OPB is critical for annotation of model components: it provides the well-understood foundation of physics for understanding biological processes – whether those processes are biochemical reactions, fluid flows, or musculoskeletal movements. In fact, the COMBINE consortium (COmputational Modeling in BIology NEtwork: https://co.mbine.org) of model builders have agreed on a set of recommended ontologies, including the OPB, for exactly this sort of model annotation (Gennari et al., 2021).

As part of a standardized approach for annotating the semantics of biosimulation models, the OPB is also a crucial ingredient for model-merging methods that leverage semantics to accelerate and automate model composition. We and others have demonstrated how physical property terms from the OPB can be used in combination with other ontology terms to precisely capture the biological meaning of a model's components and thereby provide a basis for more automated model–model integration (Beard et al., 2012; Neal et al., 2015, 2019; Shahidi et al., 2021) as well as enhanced methods for search and discovery of models in repositories (Munarko et al., 2022; Sarwar et al., 2019). As discussed in Chapter 7, we also see an opportunity to use this annotation approach for experimental and clinical data. Capturing the semantics of such data would potentially facilitate their integration with models and enhance search and discovery of the data within repositories.

The organization of physics concepts underlying the OPB, namely, the relationships between physical processes, entities, properties, and dependencies, have also proved useful as a way of organizing class hierarchies in software applications. The GUI-based applications Chalkboard (Chapters 5 and 6) and SemGen (Chapter 7) as well as the software library "libOMEXmeta" (Welsh et al., 2021) all utilize object models based on these relationships to interrelate biological concepts found in physiological models. We are also working with collaborators to apply the OPB's underlying organization in the development of biomedical reference ontologies such as the Cell Behavior Ontology (Sluka et al., 2014). Because the OPB's logical structure provides a comprehensive framework for organizing physiological concepts across physical scales and research domains, we anticipate that it will be valuable for future research efforts that focus on integrating physiological information.

Furthermore, the OPB's computable knowledge of physics also provides a foundation for the development of new computational tools for physics-based data exchange, modeling, and analysis. We envision that this knowledge will substantially augment the representative and analytical power of available biomedical ontologies for the documentation and analysis of complex biophysical systems in basic research as well as clinical research. For example, our applications for qualitative inference of causality in biological systems (see Chapters 5 and 7) represent initial efforts to leverage OPB knowledge to analyze complex biological networks. As the scope of biomedical research expands to include larger and larger biological systems, it is our hope that the OPB serves as a foundational knowledge resource for integrating, organizing, and analyzing the resulting scientific knowledge.

Our work has also been driven by the rapid growth of modeling, analysis, and simulation of physics-based biosystems. As described in Chapters 1 and 2, there are now a large number (thousands) of readily available computational models of physiology. Some of these are written in standard, readily interpreted languages, such as CellML and the Systems Biology Markup Language (SBML). Such models are particularly amenable to *annotation* with terms and concepts from the OPB; indeed, as we report in Chapter 7, there are on-going efforts to leverage these annotations for improved search and model integration.

As the number and range of biosimulation models have increased, the *Physiome* vision becomes more possible: the idea that sets of models can be combined together to produce larger, more accurate, and more comprehensive models of physiology and pathology. In turn, an effective Physiome can lead to advances in disease understanding and potentially more effective treatments. These broad goals of the Physiome motivate our development of the OPB: an essential ingredient for the Physiome is the common framework in physics and biology that is represented in the OPB.

Because we aim to support mathematical models of physiology, our focus is on the measurable properties that are used to create and validate these models. In Chapter 4, we reviewed issues of measurement and units, and in Chapter 5, we provided examples of types of biosimulation models that use and build on these measurements of biological entities. Our ontology is based on the framework of systems dynamics: an engineering approach that models the dynamics as a set of *flows* among *stocks* of entities. These flows are governed by key physical laws, notably the conservation laws and rate laws that specify how flows depend on quantities of stocks and environmental forces.

OPB – QUANTITATIVE FOUNDATIONS; SEMANTIC PERSPECTIVES

The OPB representational framework described in Chapter 6 offers both a theoretical framework and a working tool for representing and computing on concepts and rules of classical physics. It is based on a century of classical physics thought starting with Maxwell (1871) followed by formulations based on system dynamics as applied in engineering, bioengineering, and biophysics. The subsequent work of Oster et al. (1971) and Karnopp et al. (2005) formulated the "tetrahedron of state" (Figure 6.1) that recognizes the four cardinal dynamical properties and their dependencies.

Whereas the tetrahedron of state offers insights for analyzing and understanding dynamical systems, it has had no formal expression as needed for computational analysis. As discussed in Chapter 6, the OPB represents each of the four cardinal dynamical property classes (force, amount, flow rate, momentum) along with classes that represent dynamical continuity equations (see Figure 6.11), constitutive relations (Figure 6.12), and their thermodynamic properties (Figure 6.21). Figure 8.1 combines these three aspects of the OPB to extend and elaborate the tetrahedron of state to encompass essential aspects of multiscale, multidomain systems.

Diagramming the OPB property-dependency relations, as we have in Figure 8.1, exposes a remarkable symmetry wherein, for example, kinetic energy is an attribute of an occurrence of an inductive process (e.g., throwing a ball), potential energy of a capacitive process (e.g., lifting a ball), and

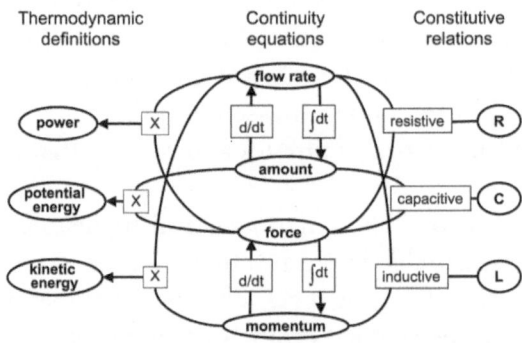

FIGURE 8.1 This diagram of the OPB schema represents the four dynamical properties (flow rate, amount, force, and momentum) (middle) along with their dependency relations including: (1) temporal integral and differential continuity dependencies (midline), (2) resistive, capacitive, inductive constitutive dependencies (right), and (3) thermodynamic property definitions for power, potential energy, and kinetic energy (left).

FIGURE 8.2 The OPB "signet ring" logo (see the book cover) is a schematic that emphasizes the symmetries of Figure 8.1.

power of a resistive process (e.g., a ball slowed by air resistance). These relations and dependencies have been commented on by others and have inspired us to fashion the "signet ring" logo in Figure 8.2 in which a four-part central axis corresponds to the four cardinal properties linked by their continuity dependencies, the loops on the right correspond to their constitutive relations, and the loops on the left to their thermodynamic properties and definitions. Lined up in the center of Figure 8.2 are the four cardinal dynamical properties: flow rate, amount, force, and momentum.

Bibliography

Ajmera, I., Swat, M., Laibe, C., Le Novère, N., & Chelliah, V. (2013). The impact of mathematical modeling on the understanding of diabetes and related complications. *CPT: Pharmacometrics & Systems Pharmacology*, *2*(7), 1–14. https://doi.org/10.1038/psp.2013.30

Alberts, B., Bray, D., Lewis, J., Raff, M., Roberts, K., & Watson, J. D. (2007). Molecular biology of the cell (4th ed.). Garland Science.

Alcántara, R., Axelsen, K. B., Morgat, A., Belda, E., Coudert, E., Bridge, A., Cao, H., De Matos, P., Ennis, M., Turner, S., Owen, G., Bougueleret, L., Xenarios, I., & Steinbeck, C. (2012). Rhea: A manually curated resource of biochemical reactions. *Nucleic Acids Research*, *40*(D1), D754–D760. https://doi.org/10.1093/NAR/GKR1126

Allemang, D., & Hendler, J. (2011). *Semantic web for the working ontologist - Effective modeling in RDFS and OWL* (2nd ed.). https://books.google.com/books/about/Semantic_Web_for_the_Working_Ontologist.html?id=_qGKPOlB1DgC

An, G., Mi, Q., Dutta-Moscato, J., & Vodovotz, Y. (2009). Agent-based models in translational systems biology. Wiley interdisciplinary reviews. *Systems Biology and Medicine*, *1*(2), 159–171. https://doi.org/10.1002/wsbm.45

Anderson, M., Meyer, B., & Olivier, P. (2002). *Diagrammatic representation and reasoning*. Springer.

Arp, R., Smith, B., & Spear, A. (2015). *Building ontologies with basic formal ontology*. The MIT Press. https://mitpress.mit.edu

Ashburner, M., Ball, C. A., Blake, J. A., Botstein, D., Butler, H., Cherry, J. M., Davis, A. P., Dolinski, K., Dwight, S. S., Eppig, J. T., Harris, M. A., Hill, D. P., Issel-Tarver, L., Kasarskis, A., Lewis, S., Matese, J. C., Richardson, J. E., Ringwald, M., Rubin, G. M., & Sherlock, G. (2000). Gene ontology: Tool for the unification of biology. *Nature Genetics*, *25*(1), 25–29. https://doi.org/10.1038/75556

Bairoch, A. (2000). The ENZYME database in 2000. *Nucleic Acids Research*, *28*(1), 304–305. https://doi.org/10.1093/NAR/28.1.304

Bansal, P., Morgat, A., Axelsen, K. B., Muthukrishnan, V., Coudert, E., Aimo, L., Hyka-Nouspikel, N., Gasteiger, E., Kerhornou, A., Neto, T. B., Pozzato, M., Blatter, M.-C., Ignatchenko, A., Redaschi, N., & Bridge, A. (2022). Rhea, the reaction knowledgebase in 2022. *Nucleic Acids Research*, *50*(D1), D693–D700. https://doi.org/10.1093/nar/gkab1016

Bassingthwaighte, J. B. (2000). Strategies for the physiome project. *Annals of Biomedical Engineering*, 28(8), 1043–1058. https://doi.org/10.1114/1.1313771/METRICS

Bassingthwaighte, J. B., Chinard, F. P., Crone, C., Goresky, C. A., Lassen, N. A., Reneman, R. S., & Zierler, K. L. (1986). Terminology for mass transport and exchange. *American Journal of Physiology*, 250(4 Pt 2), H539–H545. https://www.ncbi.nlm.nih.gov/entrez/query.fcgi?cmd=Retrieve&db=PubMed&dopt=Citation&list_uids=3963211

Bateman, A., Martin, M. J., Orchard, S., Magrane, M., Agivetova, R., Ahmad, S., Alpi, E., Bowler-Barnett, E. H., Britto, R., Bursteinas, B., Bye-A-Jee, H., Coetzee, R., Cukura, A., da Silva, A., Denny, P., Dogan, T., Ebenezer, T. G., Fan, J., Castro, L. G., ... Teodoro, D. (2021). UniProt: The universal protein knowledgebase in 2021. *Nucleic Acids Research*, 49(D1), D480–D489. https://doi.org/10.1093/NAR/GKAA1100

Beard, D. A., Liang, S. D., & Qian, H. (2002). Energy balance for analysis of complex metabolic networks. *Biophysical Journal*, 83(1), 79–86. https://www.ncbi.nlm.nih.gov/entrez/query.fcgi?cmd=Retrieve&db=PubMed&dopt=Citation&list_uids=12080101

Beard, Daniel A., & Qian, H. (2008). *Chemical Biophysics: Quantitative Analysis of Cellular Systems*. Cambridge University Press.

Beard, Daniel A., Neal, M. L., Tabesh-Saleki, N., Thompson, C. T., Bassingthwaighte, J. B., Shimoyama, M., & Carlson, B. E. (2012). Multiscale modeling and data integration in the virtual physiological rat project. *Annals of Biomedical Engineering*, 40(11), 2365–2378. https://doi.org/10.1007/s10439-012-0611-7

Bergman, R. N. (2021). Origins and history of the minimal model of glucose regulation. *Frontiers in Endocrinology*, 11(1151). https://doi.org/10.3389/fendo.2020.583016

Bergmann, F. T., Adams, R., Moodie, S., Cooper, J., Glont, M., Golebiewski, M., Hucka, M., Laibe, C., Miller, A. K., Nickerson, D. P., Olivier, B. G., Rodriguez, N., Sauro, H. M., Scharm, M., Soiland-Reyes, S., Waltemath, D., Yvon, F., & Le Novère, N. (2014). COMBINE archive and OMEX format: One file to share all information to reproduce a modeling project. *BMC Bioinformatics*, 15, 369. https://doi.org/10.1186/s12859-014-0369-z

Berners-Lee, T., Hendler, J., & Lassila, O. (2001). The semantic web. *Scientific American*, 284(5), 34–43. https://www.jstor.org/stable/26059207

Blake, J. A., Christie, K. R., Dolan, M. E., Drabkin, H. J., Hill, D. P., Ni, L., Sitnikov, D., Burgess, S., Buza, T., Gresham, C., McCarthy, F., Pillai, L., Wang, H., Carbon, S., Dietze, H., Lewis, S. E., Mungall, C. J., Munoz-Torres, M. C., Feuermann, M., ... Westerfeld, M. (2015). Gene Ontology Consortium: Going forward. *Nucleic Acids Research*, 43(D1), D1049–D1056. https://doi.org/10.1093/NAR/GKU1179

Blinov, M. L., Schaff, J. C., Vasilescu, D., Moraru, I. I., Bloom, J. E., & Loew, L. M. (2017). Compartmental and spatial rule-based modeling with virtual cell. *Biophysical Journal*, 113(7), 1365–1372. https://doi.org/10.1016/J.BPJ.2017.08.022

Bock, C., & Gruninger, M. (2005). PSL: A semantic domain for flow models. *Software and Systems Modeling, 4,* 209–231. https://doi.org/10.1007/s10270-004-0066-x

Borgo, S., & Masolo, C. (2010). Ontological foundations of DOLCE. *Theory and Applications of Ontology: Computer Applications,* 279–295. https://doi.org/10.1007/978-90-481-8847-5_13/COVER

Borst, P., Akkermans, H., & Top, J. (1997). Engineering ontologies. *International Journal of Human-Computer Studies, 46,* 365–406.

Borst, P., Akkermans, J., Pos, A., & Top, J. (1995). The PhysSys ontology for physical systems. In B. Bredeweg (Ed.), *Working Papers of the Ninth International Workshop on Qualitative Reasoning QR'95* (pp. 11–21). University of Amsterdam. https://www.qrg.northwestern.edu/papers/Files/qr-workshops/QR95/Borst_1995_PhysSys_Ontology_Physical_Systems.pdf

Canetea, J., Pimentel, V., Barbancho, J., & Luque, A. (2019). System dynamics modelling approach in Health Sciences. Application to the regulation of the cardiovascular function. *Informatics in Medicine Unlocked, 15.* https://doi.org/https://doi.org/10.1016/j.imu.2019.100164

Carbon, S., Douglass, E., Dunn, N., Good, B., Harris, N. L., Lewis, S. E., Mungall, C. J., Basu, S., Chisholm, R. L., Dodson, R. J., Hartline, E., Fey, P., Thomas, P. D., Albou, L. P., Ebert, D., Kesling, M. J., Mi, H., Muruganujan, A., Huang, X., ... Westerfield, M. (2019). The Gene Ontology resource: 20 years and still going strong. *Nucleic Acids Research, 47*(D1), D330–D338. https://doi.org/10.1093/NAR/GKY1055

Catlett, N. L., Bargnesi, A. J., Ungerer, S., Seagaran, T., Ladd, W., Elliston, K. O., & Pratt, D. (2013). Reverse causal reasoning: Applying qualitative causal knowledge to the interpretation of high-throughput data. *BMC Bioinformatics, 14*(1), 1–14. https://doi.org/10.1186/1471-2105-14-340/FIGURES/4

Cheong, H., & Butscher, A. (2019). Physics-based simulation ontology: An ontology to support modelling and reuse of data for physics-based simulation. *Journal of Engineering Design, 30*(10–12), 655–687. https://doi.org/10.1080/09544828.2019.1644301

Christianson, S. (2012). *100 diagrams that changed the world.* Plume Books, Penguin Group.

Collins, J. B. (2004). Standardizing an Ontology of Physics for Modeling and Simulation. In *Fall Simulation Interoperability Workshop,* by Naval Research Lab, Washington, DC. https://apps.dtic.mil/sti/citations/ADA610086

Cook, D. L., Bookstein, F. L., & Gennari, J. H. (2011). Physical properties of biological entities: An introduction to the ontology of physics for biology. *PLoS One, 6*(12), e28708. https://doi.org/10.1371/journal.pone.0028708

Cook, D. L., Farley, J. F., & Tapscott, S. J. (2001). A basis for a visual language for describing, archiving and analyzing functional models of complex biological systems. *Genome Biology, 2*(4), RESEARCH0012. https://www.ncbi.nlm.nih.gov/entrez/query.fcgi?cmd=Retrieve&db=PubMed&dopt=Citation&list_uids=11305940

Cook, D. L., Gennari, J. H., & Neal, M. L. (2019). The Ontology of Physics for Biology - A Companion to Basic Formal Ontology. In Barry Smith (Ed.), *10th International Conference on Biomedical Ontology.*

Cook, D. L., Gennari, J. H., & Wiley, J. C. (2007). Chalkboard: Ontology-based pathway modeling and qualitative inference of disease mechanisms. *Pacific Symposium on Biocomputing, 12,* 16–27. https://psb.stanford.edu/psb-online/proceedings/psb07/cook.pdf

Cook, D. L., Neal, M. L., Hoehndorf, R., Gkoutos, G. V, & Gennari, J. H. (2013). Representing physiological processes and their participants with PhysioMaps. *Journal of Biomedical Semantics, 4*(Suppl 1), S2. https://doi.org/10.1186/2041-1480-4-S1-S2

Cook, D. L., Porte Jr., D., & Crill, W. E. (1981). Voltage dependence of rhythmic plateau potentials of pancreatic islet cells. *American Journal of Physiology, 240*(3), E290–E296.

Crampin, E. J., Smith, N. P., & Hunter, P. J. (2004). Multi-scale modelling and the IUPS physiome project. *Journal of Molecular Histology, 35*(7), 707–714. https://www.ncbi.nlm.nih.gov/entrez/query.fcgi?cmd=Retrieve&db=PubMed&dopt=Citation&list_uids=15614626

Degtyarenko, K., de Matos, P., Ennis, M., Hastings, J., Zbinden, M., McNaught, A., Alcantara, R., Darsow, M., Guedj, M., & Ashburner, M. (2008). ChEBI: A database and ontology for chemical entities of biological interest. *Nucleic Acids Res, 36*(Database issue), D344-50. https://www.ncbi.nlm.nih.gov/entrez/query.fcgi?cmd=Retrieve&db=PubMed&dopt=Citation&list_uids=17932057

Demir, E., Cary, M. P., Paley, S., Fukuda, K., Lemer, C., Vastrik, I., Wu, G., D'Eustachio, P., Schaefer, C., Luciano, J., Schacherer, F., Martinez-Flores, I., Hu, Z., Jimenez-Jacinto, V., Joshi-Tope, G., Kandasamy, K., Lopez-Fuentes, A. C., Mi, H., Pichler, E., ... Bader, G. D. (2010). The BioPAX community standard for pathway data sharing. *Nature Biotechnology, 28*(9), 935–942. https://doi.org/10.1038/nbt.1666

Diehl, A. D., Meehan, T. F., Bradford, Y. M., Brush, M. H., Dahdul, W. M., Dougall, D. S., He, Y., Osumi-Sutherland, D., Ruttenberg, A., Sarntivijai, S., Van Slyke, C. E., Vasilevsky, N. A., Haendel, M. A., Blake, J. A., & Mungall, C. J. (2016). The cell ontology 2016: Enhanced content, modularization, and ontology interoperability. *Journal of Biomedical Semantics, 7*(1), 1–10. https://doi.org/10.1186/S13326-016-0088-7/TABLES/1

Duque-Ramos, A., Fernandez-Breis, J., Stevens, R., & Aussenac-Gilles, N. (2011). OQuaRE: A SQuaRE-based approach for evaluating the quality of ontologies. *Journal of Research and Practice in Information Technology, 43,* 159–176.

Erson, E. Z., & Cavusoglu, M. C. (2007). Ontology based design for integrative simulation of human physiology. *Proceedings of the International Symposium on Health Informatics and Bioinformatics,* Antalya.

Fahey, M. E., Bennett, M. J., Mahon, C., Jäger, S., Pache, L., Kumar, D., Shapiro, A., Rao, K., Chanda, S. K., Craik, C. S., Frankel, A. D., & Krogan, N. J. (2011). GPS-Prot: A web-based visualization platform for integrating host-pathogen interaction data. *BMC Bioinformatics, 12*(1), 1–13. https://doi.org/10.1186/1471-2105-12-298/FIGURES/5

Fernandez, M., Gomez-Perez, A., & Juristo, N. (1997). Methontology: From ontological art towards ontological engineering. *Proceedings AAAI Spring Symposium.* https://oa.upm.es/5484/?trk=public_post_main-feed-card-text

Feynman, R. (1994). *The character of physical law*. Modern Library.

Gawthrop, P., & Crampin, E. J. (2018). Bond graph representation of chemical reaction networks. *IEEE Transactions on NanoBioscience, 17*(4), 449–455. https://doi.org/10.1109/TNB.2018.2876391

Gawthrop, P. J., & Crampin, E. J. (2014). Energy-based analysis of biochemical cycles using bond graphs. *Proceedings: Mathematical, Physical and Engineering Sciences, 470*(2171), 20140459. https://doi.org/10.1098/rspa.2014.0459

Gennari, J. H., König, M., Misirli, G., Neal, M. L., Nickerson, D. P., & Waltemath, D. (2021). OMEX metadata specification (version 1.2). *Journal of Integrative Bioinformatics, 18*(3). https://doi.org/10.1515/JIB-2021-0020

Gennari, J. H., Neal, M. L., Galdzicki, M., & Cook, D. L. (2011). Multiple ontologies in action: Composite annotations for biosimulation models. *Journal of Biomedical Informatics, 44*(1), 146–154. https://doi.org/10.1016/j.jbi.2010.06.007

Gillespie, M., Jassal, B., Stephan, R., Milacic, M., Rothfels, K., Senff-Ribeiro, A., Griss, J., Sevilla, C., Matthews, L., Gong, C., Deng, C., Varusai, T., Ragueneau, E., Haider, Y., May, B., Shamovsky, V., Weiser, J., Brunson, T., Sanati, N., ... D'Eustachio, P. (2022). The reactome pathway knowledgebase 2022. *Nucleic Acids Research, 50*(D1), D687–D692. https://doi.org/10.1093/NAR/GKAB1028

Gkoutos, G. V., Green, E. C. J., Mallon, A. M., Hancock, J. M., & Davidson, D. (2005). Using ontologies to describe mouse phenotypes. *Genome Biology, 6*(1), 1–10. https://doi.org/10.1186/GB-2004-6-1-R8/FIGURES/4

Gkoutos, G. V., Schofield, P. N., & Hoehndorf, R. (2012). The units ontology: A tool for integrating units of measurement in science. *Database (Oxford), 2012*, bas033. https://doi.org/10.1093/database/bas033

Glimm, B., Horrocks, I., Motik, B., Stoilos, G., & Wang, Z. (2014). HermiT: An OWL 2 Reasoner. *Journal of Automated Reasoning, 53*(3), 245–269. https://doi.org/10.1007/S10817-014-9305-1/METRICS

Goldberg, R. N., Tewari, Y. B., & Bhat, T. N. (2006). *Thermodynamics of Enzyme-Catalyzed Reactions*. https://xpdb.nist.gov/enzyme_thermodynamics/enzyme_introduction.html

Gremse, M., Chang, A., Schomburg, I., Grote, A., Scheer, M., Ebeling, C., & Schomburg, D. (2011). The BRENDA Tissue Ontology (BTO): The first all-integrating ontology of all organisms for enzyme sources. *Nucleic Acids Research, 39*(suppl_1), D507–D513. https://doi.org/10.1093/NAR/GKQ968

Gruber, T. R. (1995). Toward principles for the design of ontologies used for knowledge sharing. *International Journal of Human-Computer Studies, 43*(5–6), 907–928.

Gruber, T. R, & Olsen, G. R. (1994). An Ontology for Engineering Mathematics. In J. Doyle, P. Torasso, & E. Sandewall (Eds.), *Fourth International Conference on Principles of Knowledge Representation and Reasoning*. Morgan Kaufmann. https://www-ksl.stanford.edu/knowledge-sharing/papers/engmath.html

Guarino, N., & Welty, C. (2002). Evaluating ontological decisions with OntoClean. *Communications of the ACM, 45*(2), 61–65. https://doi.org/10.1145/503124.503150

Gündel, M., Younesi, E., Malhotra, A., Wang, J., Li, H., Zhang, B., de Bono, B., Mevissen, H. T., & Hofmann-Apitius, M. (2013). HuPSON: The human physiology simulation ontology. *Journal of Biomedical Semantics*, 4(1), 1–9. https://doi.org/10.1186/2041-1480-4-35/TABLES/3

Hastings, J., Owen, G., Dekker, A., Ennis, M., Kale, N., Muthukrishnan, V., Turner, S., Swainston, N., Mendes, P., & Steinbeck, C. (2016). ChEBI in 2016: Improved services and an expanding collection of metabolites. *Nucleic Acids Research*, 44(D1), D1214–D1219. https://doi.org/10.1093/NAR/GKV1031

Hayamizu, T. F., Mangan, M., Corradi, J. P., Kadin, J. A., & Ringwald, M. (2005). The Adult Mouse Anatomical Dictionary: A tool for annotating and integrating data. *Genome Biology*, 6(3), 1–8. https://doi.org/10.1186/GB-2005-6-3-R29/COMMENTS

Hayamizu, T. F., Wicks, M. N., Davidson, D. R., Burger, A., Ringwald, M., & Baldock, R. A. (2013). EMAP/EMAPA ontology of mouse developmental anatomy: 2013 update. *Journal of Biomedical Semantics*, 4(1), 1–5. https://doi.org/10.1186/2041-1480-4-15/FIGURES/2

Hemker, H. C., & Beguin, S. (1993). Standard and method independent units for heparin anticoagulant activities. *Thrombosis and Haemostasis*, 70(5), 724–728. https://www.ncbi.nlm.nih.gov/pubmed/8128425

Hendler, J. (2001). Agents and the semantic web. *IEEE Intelligent Systems*, 16(2), 30–37. https://doi.org/10.1109/5254.920597

Herre, H. (2010). General Formal Ontology (GFO): A Foundational Ontology for Conceptual Modelling. In R. Poli, M. Healy, A. Kameas (Eds.), *Theory and applications of ontology: Computer applications*. https://doi.org/10.1007/978-90-481-8847-5_14

Hille, B. (2001). *Ion channels of excitable membranes* (3rd ed.). Sinauer Associates, Inc.

Horrocks, I., Patel-Schneider, P., Boley, H., Tabet, S., Grosof, B., & Dean, M. (n.d.). *SWRL: A Semantic Web Rule Language Combining OWL and RuleML*. Retrieved May 22, 2023, from https://www.w3.org/Submission/SWRL/

Hucka, M., Finney, A., Sauro, H. M., Bolouri, H., Doyle, J. C., Kitano, H., Arkin, A. P., Bornstein, B. J., Bray, D., Cornish-Bowden, A., ... SBML Forum. (2003). The systems biology markup language (SBML): A medium for representation and exchange of biochemical network models. *Bioinformatics*, 19(4), 524–531. https://doi.org/10.1093/BIOINFORMATICS/BTG015

Hunter, P. J., & Borg, T. K. (2003). Integration from proteins to organs: The Physiome Project. *Nature Reviews Molecular Cell Biology 2003*, 4(3), 237–243. https://doi.org/10.1038/nrm1054

Husakova, M. (2015). *Conceptual modelling in computational immunology*. Tomas Bruckner, Repin-Zivonin. https://pub.bruckner.cz

Jensen, K., & Kristensen, L. M. (2009). *Coloured petri nets: Modelling and validation of concurrent systems*. Springer.

Kanehisa, M., Furumichi, M., Sato, Y., Kawashima, M., & Ishiguro-Watanabe, M. (2023). KEGG for taxonomy-based analysis of pathways and genomes. *Nucleic Acids Research*, 51(D1), D587–D592. https://doi.org/10.1093/NAR/GKAC963

Kanehisa, M., & Goto, S. (2000). KEGG: Kyoto encyclopedia of genes and genomes. *Nucleic Acids Research*, 28(1), 27–30. https://doi.org/10.1093/NAR/28.1.27

Karnopp, D. (1979). *Bond graph techniques for dynamic systems in engineering and biology*. Pergamon Press.

Karnopp, D., Margolis, D. L., & Rosenberg, R. C. (2005). *System dynamics: Modeling and simulation of mechanotronic systems* (4th ed.). John Wiley & Sons.

Karp, P. D., Billington, R., Caspi, R., Fulcher, C. A., Latendresse, M., Kothari, A., Keseler, I. M., Krummenacker, M., Midford, P. E., Ong, Q., Ong, W. K., Paley, S. M., & Subhraveti, P. (2019). The BioCyc collection of microbial genomes and metabolic pathways. *Briefings in Bioinformatics, 20*(4), 1085–1093. https://doi.org/10.1093/BIB/BBX085

Karr, J. R., Sanghvi, J. C., Macklin, D. N., Arora, A., & Covert, M. W. (2013). WholeCellKB: Model organism databases for comprehensive whole-cell models. *Nucleic Acids Research, 41*(Database issue), D787–D792. https://doi.org/10.1093/nar/gks1108

Katchalsky, A., & Kedemo. (1962). Thermodynamics of flow processes in biological systems. *Biophysical Journal, 2*(2)Pt2, 53–78. https://www.ncbi.nlm.nih.gov/entrez/query.fcgi?cmd=Retrieve&db=PubMed&dopt=Citation&list_uids=14454230

Kauffman, S. A. (1993). *The origins of order : Self-organization and selection in evolution*. Oxford University Press, Inc.

Kerckhoffs, R. C., Neal, M. L., Gu, Q., Bassingthwaighte, J. B., Omens, J. H., & McCulloch, A. D. (2007). Coupling of a 3D finite element model of cardiac ventricular mechanics to lumped systems models of the systemic and pulmonic circulation. *Annals of Biomedical Engineering, 35*(1), 1–18. https://www.ncbi.nlm.nih.gov/entrez/query.fcgi?cmd=Retrieve&db=PubMed&dopt=Citation&list_uids=17111210

King, Z. A., Lu, J., Dräger, A., Miller, P., Federowicz, S., Lerman, J. A., Ebrahim, A., Palsson, B. O., & Lewis, N. E. (2016). BiGG models: A platform for integrating, standardizing and sharing genome-scale models. *Nucleic Acids Research, 44*(D1), D515–D522. https://doi.org/10.1093/nar/gkv1049

Köhler, S., Gargano, M., Matentzoglu, N., Carmody, L. C., Lewis-Smith, D., Vasilevsky, N. A., Danis, D., Balagura, G., Baynam, G., Brower, A. M., Callahan, T. J., Chute, C. G., Est, J. L., Galer, P. D., Ganesan, S., Griese, M., Haimel, M., Pazmandi, J., Hanauer, M., ... Robinson, P. N. (2021). The human phenotype ontology in 2021. *Nucleic Acids Research, 49*(D1), D1207–D1217. https://doi.org/10.1093/NAR/GKAA1043

Kosko, B. (1993). *Fuzzy thinking: The new science of fuzzy logic*. Hyperion.

Krogh, A. (2008). What are artificial neural networks? *Nature Biotechnology 2008, 26*(2), 195–197. https://doi.org/10.1038/NBT1386

Le Novere, N., Finney, A., Hucka, M., Bhalla, U. S., Campagne, F., Collado-Vides, J., Crampin, E. J., Halstead, M., Klipp, E., Mendes, P., Nielsen, P., Sauro, H., Shapiro, B., Snoep, J. L., Spence, H. D., & Wanner, B. L. (2005). Minimum information requested in the annotation of biochemical models (MIRIAM). *Nature Biotechnology, 23*(12), 1509–1515. https://www.ncbi.nlm.nih.gov/entrez/query.fcgi?cmd=Retrieve&db=PubMed&dopt=Citation&list_uids=16333295

Le Rolle, V., Hernandez, A. I., Richard, P. Y., Buisson, J., & Carrault, G. (2005). A bond graph model of the cardiovascular system. *Acta Biotheoretica, 53*(4), 295–312. https://www.ncbi.nlm.nih.gov/entrez/query.fcgi?cmd=Retrieve&d b=PubMed&dopt=Citation&list_uids=16583271

Lefèvre, J., Lefèvre, L., & Couteiro, B. (1999). A bond graph model of chemo-mechanical transduction in the mammalian left ventricle. *Simulation Practice and Theory, 7*(5–6), 531–552. https://doi.org/10.1016/ S0928-4869(99)00023-3

Lehmann, H. P., Fuentes-Arderiu, X., & Bertello, L. F. (1996). Glossary of terms in quantities and units in clinical chemistry. *Pure and Applied Chemistry, 68*(4), 957–1000.

Lehne, B., & Schlitt, T. (2009). Protein-protein interaction databases: Keeping up with growing interactomes. *Human Genomics, 3*(3), 291–297. https://doi. org/10.1186/1479-7364-3-3-291/TABLES/3

Lindberg, D. A. B., Humphreys, B. L., & McCray, A. T. (1993). The unified medical language system. *Methods of Information in Medicine, 32*(4), 281–291. https://doi.org/10.1055/S-0038-1634945/ID/BR1634945-20

Lloyd, C. M., Halstead, M. D. B., & Nielsen, P. F. (2004). CellML: Its future, present and past. *Progress in Biophysics and Molecular Biology, 85*(2–3), 433–450. https://doi.org/10.1016/J.PBIOMOLBIO.2004.01.004

Lord, P., & Stevens, R. (2010). Adding a little reality to building ontologies for biology. *PLoS ONE, 5*(9), e12258. https://www.ncbi.nlm.nih.gov/entrez/query. fcgi?cmd=Retrieve&db=PubMed&dopt=Citation&list_uids=20838431

Malik-Sheriff, R. S., Glont, M., Nguyen, T. V. N., Tiwari, K., Roberts, M. G., Xavier, A., Vu, M. T., Men, J., Maire, M., Kananathan, S., Fairbanks, E. L., Meyer, J. P., Arankalle, C., Varusai, T. M., Knight-Schrijver, V., Li, L., Dueñas-Roca, C., Dass, G., Keating, S. M., ... Hermjakob, H. (2020). BioModels-15 years of sharing computational models in life science. *Nucleic Acids Research, 48*(D1), D407–D415. https://doi.org/10.1093/NAR/GKZ1055

Margolis, D. L. (1985). A survey of bond graph modeling for interacting lumped and distributed systems. *Franklin Institute Journal, 319*, 125–135. https:// md1.csa.com/partners/viewrecord.php?requester=gs&collection=TRD&reci d=A8530854AH

Mari, A., Tura, A., Grespan, E., & Bizzotto, R. (2020). Mathematical modeling for the physiological and clinical investigation of glucose homeostasis and diabetes. *Frontiers in Physiology, 11*. https://doi.org/10.3389/fphys.2020.575789

Maxwell, J. C. (1871). Remarks on the mathematical classification of physical quantities. *Proceedings of the London Mathematical Society, 2*(34), 224–232.

Meehan, T. F., Masci, A. M., Abdulla, A., Cowell, L. G., Blake, J. A., Mungall, C. J., & Diehl, A. D. (2011). Logical development of the cell ontology. *BMC Bioinformatics, 12*(1), 1–12. https://doi.org/10.1186/1471-2105-12-6/ FIGURES/6

Meldal, B. H. M., Bye-A-Jee, H., Gajdoš, L., Hammerová, Z., Horáčková, A., Melicher, F., Perfetto, L., Pokorný, D., Lopez, M. R., Türková, A., Wong, E. D., Xie, Z., Casanova, E. B., Del-Toro, N., Koch, M., Porras, P., Hermjakob, H.,

& Orchard, S. (2019). Complex portal 2018: Extended content and enhanced visualization tools for macromolecular complexes. *Nucleic Acids Research*, *47*(D1), D550–D558. https://doi.org/10.1093/NAR/GKY1001

Michaelis, L., Menten, M. L., Johnson, K. A., & Goody, R. S. (2011). The original Michaelis constant: Translation of the 1913 Michaelis-Menten paper. *Biochemistry*, *50*(39), 8264–8269. https://doi.org/10.1021/bi201284u

Mikulecky, D. C. (1983). Network thermodynamics: A candidate for a common language for theoretical and experimental biology. *American Journal of Physiology*, *245*(1), R1–R9. https://www.ncbi.nlm.nih.gov/entrez/query.fcgi?cmd=Retrieve&db=PubMed&dopt=Citation&list_uids=6869569

Mirzadeh, Z., Faber, C. L., & Schwartz, M. W. (2022). Central Nervous system control of glucose homeostasis: A therapeutic target for type 2 diabetes? Annual Review of Pharmacology and Toxicology, *62*, 55–84. https://doi.org/doi/10.1146/annurev-pharmtox-052220-010446

Munarko, Y., Sarwar, D. M., Rampadarath, A., Atalag, K., Gennari, J. H., Neal, M. L., & Nickerson, D. P. (2022). NLIMED: Natural language interface for model entity discovery in biosimulation model repositories. *Frontiers in Physiology*, *13*. https://www.frontiersin.org/articles/10.3389/fphys.2022.820683

Mungall, C. J., Dietze, H., & Osumi-Sutherland, D. (2014). Use of OWL within the Gene Ontology. *BioRxiv*, 010090. https://doi.org/10.1101/010090

Mungall, C. J., Torniai, C., Gkoutos, G. V., Lewis, S. E., & Haendel, M. A. (2012). Uberon, an integrative multi-species anatomy ontology. *Genome Biology*, *13*(1), 1–20. https://doi.org/10.1186/GB-2012-13-1-R5/FIGURES/5

Natale, D. A., Arighi, C. N., Blake, J. A., Bona, J., Chen, C., Chen, S. C., Christie, K. R., Cowart, J., D'Eustachio, P., Diehl, A. D., Drabkin, H. J., Duncan, W. D., Huang, H., Ren, J., Ross, K., Ruttenberg, A., Shamovsky, V., Smith, B., Wang, Q., ... Wu, C. H. (2017). Protein Ontology (PRO): Enhancing and scaling up the representation of protein entities. *Nucleic Acids Research*, *45*(D1), D339–D346. https://doi.org/10.1093/NAR/GKW1075

Neal, M. L., Gennari, J. H., & Cook, D. L. (2016). Qualitative causal analyses of biosimulation models. *International Conference on Biomedical Ontology and BioCreative (ICBO BioCreative 2016)*, *Vol-1747*|u(IT604). https://ceur-ws.org/Vol-1747/IT604_ICBO2016.pdf

Neal, M. L., Carlson, B. E., Thompson, C. T., James, R. C., Kim, K. G., Tran, K., Crampin, E. J., Cook, D. L., & Gennari, J. H. (2015). Semantics-based composition of integrated cardiomyocyte models motivated by real-world use cases. *PLoS One*, *10*(12), e0145621. https://doi.org/10.1371/journal.pone.0145621

Neal, M. L., Cooling, M. T., Smith, L. P., Thompson, C. T., Sauro, H. M., Carlson, B. E., Cook, D. L., & Gennari, J. H. (2014). A reappraisal of how to build modular, reusable models of biological systems. *PLOS Computational Biology*, *10*(10). https://doi.org/10.1371/journal.pcbi.1003849

Neal, M. L., Thompson, C. T., Kim, K. G., James, R. C., Cook, D. L., Carlson, B. E., & Gennari, J. H. (2019a). SemGen: A tool for semantics-based annotation and composition of biosimulation models. *Bioinformatics*, *35*(9), 1600–1602. https://doi.org/10.1093/bioinformatics/bty829

Neal, M. L., König, M., Nickerson, D., Misirli, G., Kalbasi, R., Dräger, A., Atalag, K., Chelliah, V., Cooling, M. T., Cook, D. L., Crook, S., De Alba, M., Friedman, S. H., Garny, A., Gennari, J. H., Gleeson, P., Golebiewski, M., Hucka, M., Juty, N., … Waltemath, D. (2019b). Harmonizing semantic annotations for computational models in biology. *Briefings in Bioinformatics*, *20*(2), 540–550. https://doi.org/10.1093/bib/bby087

Nickerson, D. P., Atalag, K., Bono, B. de, Geiger, J., Goble, C., Hollmann, S., Lonien, J., Mueller, W., Regierer, B., Stanford, N. J., Golebiewski, M., & Hunter, P. (2016). The Human Physiome: How standards, software and innovative service infrastructures are providing the building blocks to make it achievable. *Interface Focus*, 6, 20150103. https://doi.org/10.1098/rsfs.2015.0103

Novère, N. Le, Hucka, M., Mi, H., Moodie, S., Schreiber, F., Sorokin, A., Demir, E., Wegner, K., Aladjem, M. I., Wimalaratne, S. M., Bergman, F. T., Gauges, R., Ghazal, P., Kawaji, H., Li, L., Matsuoka, Y., Villéger, A., Boyd, S. E., Calzone, L., … Kitano, H. (2009). The systems biology graphical notation. *Nature Biotechnology*, *27*(8), 735–741. https://doi.org/10.1038/nbt.1558

Noy, N. F., & McGuinness, D. L. (2001). *Ontology Development 101*. Knowledge Systems Laboratory, Stanford University.

Noy, N. F., Shah, N. H., Whetzel, P. L., Dai, B., Dorf, M., Griffith, N., Jonquet, C., Rubin, D. L., Storey, M. A., Chute, C. G., & Musen, M. A. (2009). BioPortal: Ontologies and integrated data resources at the click of a mouse. *Nucleic Acids Research*, *37*(Web Server issue), W170-3. https://www.ncbi.nlm.nih.gov/entrez/query.fcgi?cmd=Retrieve&db=PubMed&dopt=Citation&list_uids=19483092

Orii, N., & Ganapathiraju, M. K. (2012). Wiki-Pi: A web-server of annotated human protein-protein interactions to aid in discovery of protein function. *PLoS One*, *7*(11), e49029. https://doi.org/10.1371/JOURNAL.PONE.0049029

Oster, G. F., Perelson, A. S., & Katchalsky, A. (1971). Network thermodynamics. *Nature*, *234*(5329), 393–399.

Oster, G. F., Perelson, A. S., & Katchalsky, A. (1973). Network thermodynamics: Dynamic modelling of biophysical systems. *Quarterly Reviews of Biophysics*, *6*(1), 1–134. https://www.ncbi.nlm.nih.gov/entrez/query.fcgi?cmd=Retrieve&db=PubMed&dopt=Citation&list_uids=4576440

Ozgovde, A., & Gruninger, M. (2010). Foundational Process Relations in Bio-Ontologies. In A. Galton & R. Mizoguchi (Eds.), *Formal ontology in information systems*. IOS Press. https://doi.org/10.3233/978-1-60750-535-8-243

Peleg, M., Tu, S., Manindroo, A., & Altman, R. B. (2004). Modeling and analyzing biomedical processes using workflow/Petri Net models and tools. *MedInfo*, *11*(Pt 1), 74–78. https://www.ncbi.nlm.nih.gov/entrez/query.fcgi?cmd=Retrieve&db=PubMed&dopt=Citation&list_uids=15360778

Perelson, A. S. (1975). Network thermodynamics. An overview. *Biophysical Journal*, *15*(7), 667–685. https://www.ncbi.nlm.nih.gov/entrez/query.fcgi?cmd=Retrieve&db=PubMed&dopt=Citation&list_uids=1095093

Qian, H., & Beard, D. A. (2005). Thermodynamics of stoichiometric biochemical networks in living systems far from equilibrium. *Biophysical Chemistry*, *114*(2–3), 213–220. https://www.ncbi.nlm.nih.gov/entrez/query.fcgi?cmd=Retrieve&db=PubMed&dopt=Citation&list_uids=15829355

Robinson, P. N., Köhler, S., Bauer, S., Seelow, D., Horn, D., & Mundlos, S. (2008). The human phenotype ontology: A tool for annotating and analyzing human hereditary disease. *The American Journal of Human Genetics*, *83*(5), 610–615. https://doi.org/10.1016/j.ajhg.2008.09.017

Rosse, C., Ben Said, M., Eno, K. R., & Brinkley, J. F. (1995). Enhancements of anatomical information in UMLS knowledge sources. *Proceedings of the Annual Symposium on Computer Applications in Medical Care*, 873–877. https://www.ncbi.nlm.nih.gov/entrez/query.fcgi?cmd=Retrieve&db=PubMed&dopt=Citation&list_uids=8563417

Rosse, C., Mejino, J. L., Modayur, B. R., Jakobovits, R., Hinshaw, K. P., & Brinkley, J. F. (1998). Motivation and organizational principles for anatomical knowledge representation: The digital anatomist symbolic knowledge base. *Journal of the American Medical Informatics Association*, *5*(1), 17–40. https://www.ncbi.nlm.nih.gov/entrez/query.fcgi?cmd=Retrieve&db=PubMed&dopt=Citation&list_uids=9452983

Rosse, C., & Mejino, J. L. V. (2003). A reference ontology for biomedical informatics: The Foundational Model of Anatomy. *Journal of Biomedical Informatics*, *36*(6), 478–500. https://doi.org/10.1016/J.JBI.2003.11.007

Rosse, C., Shapiro, L. G., & Brinkley, J. F. (1998). The digital anatomist foundational model: Principles for defining and structuring its concept domain. *Proceedings. AMIA Symposium*, 820–824. https://www.ncbi.nlm.nih.gov/entrez/query.fcgi?cmd=Retrieve&db=PubMed&dopt=Citation&list_uids=9929333

Rovelli, C. (2018). *The order of time*. Penguin Random House.

Rzhetsky, A., & Evans, J. A. (2011). War of ontology worlds: Mathematics, computer code, or esperanto? *PLoS Computational Biology*, *7*(9), e1002191.

Saito, R., Smoot, M. E., Ono, K., Ruscheinski, J., Wang, P. L., Lotia, S., Pico, A. R., Bader, G. D., & Ideker, T. (2012). A travel guide to Cytoscape plugins. *Nature Methods*, *9*(11), 1069–1076. https://doi.org/10.1038/nmeth.2212

Sakmann, B., & Neher, E. (1983). *Single-channel recording*. Plenum Publishing Corp.

Sarwar, D. M., Kalbasi, R., Gennari, J. H., Carlson, B. E., Neal, M. L., Bono, B. de, Atalag, K., Hunter, P. J., & Nickerson, D. P. (2019). Model annotation and discovery with the Physiome Model Repository. *BMC Bioinformatics*, *20*(1), 457. https://doi.org/10.1186/s12859-019-2987-y

Schadow, G., McDonald, C. J., Suico, J. G., Fohring, U., & Tolxdorff, T. (1999). Units of measure in clinical information systems. *Journal of the American Medical Informatics Association*, *6*(2), 151–162. https://www.ncbi.nlm.nih.gov/pmc/articles/PMC61354/pdf/0060151.pdf

Segel, I. H. (1975). *Enzyme kinetics: Behavior and analysis of rapid equilibrium and steady-state enzyme systems*. Wiley-Interscience Publication.

Shahidi, N., Pan, M., Safaei, S., Tran, K., Crampin, E. J., & Nickerson, D. P. (2021). Hierarchical semantic composition of biosimulation models using bond graphs. *PLoS Computational Biology*, *17*(5). https://doi.org/10.1371/JOURNAL.PCBI.1008859

Shannon, P. (2003). Cytoscape: A software environment for integrated models of biomolecular interaction networks. *Genome Research*, *13*, 2498–2504.

Sluka, J. P., Shirinifard, A., Swat, M., Cosmanescu, A., Heiland, R. W., & Glazier, J. A. (2014). The cell behavior ontology: Describing the intrinsic biological behaviors of real and model cells seen as active agents. *Bioinformatics*, *30*(16), 2367–2374. https://doi.org/10.1093/bioinformatics/btu210

Smith, B., Ashburner, M., Rosse, C., Bard, J., Bug, W., Ceusters, W., Goldberg, L. J., Eilbeck, K., Ireland, A., Mungall, C. J., Leontis, N., Rocca-Serra, P., Ruttenberg, A., Sansone, S. A., Scheuermann, R. H., Shah, N., Whetzel, P. L., & Lewis, S. (2007). The OBO Foundry: Coordinated evolution of ontologies to support biomedical data integration. *Nature Biotechnology*, *25*(11), 1251–1255. https://www.ncbi.nlm.nih.gov/entrez/query.fcgi?cmd=Retrieve&db=PubMed&dopt=Citation&list_uids=17989687

Smith, B., Ceusters, W., Klagges, B., Kohler, J., Kumar, A., Lomax, J., Mungall, C., Neuhaus, F., Rector, A. L., & Rosse, C. (2005). Relations in biomedical ontologies. *Genome Biology*, *6*(5), R46. https://www.ncbi.nlm.nih.gov/entrez/query.fcgi?cmd=Retrieve&db=PubMed&dopt=Citation&list_uids=15892874

Smith, B., Lewis, S., & Ashburner, M. (Eds.). (2006). *OBO Foundry: A New Paradigm for Biomedical Ontology Development*. https://obofoundry.org/

Smith, Barry. (2013). Classifying Processes: An Essay in Applied Ontology. In D. S. Oderberg (Ed.), *Classifying reality* (pp. 101–126). John Wiley & Sons, Ltd. https://doi.org/10.1002/9781118627747.CH6

Steil, G. M., Rebrin, K., Darwin, C., Hariri, F., & Saad, M. F. (2006). Feasibility of automating insulin delivery for the treatment of type 1 diabetes. *Diabetes*, *55*(12), 3344. https://doi.org/10.2337/db06-0419

Strogatz, S. H. (2019). *Infinite powers. How calculus reveals the secrets of the universe*. Houghton, Mifflin, Harcourt.

Thomas, P. D., Hill, D. P., Mi, H., Osumi-Sutherland, D., Van Auken, K., Carbon, S., Balhoff, J. P., Albou, L. P., Good, B., Gaudet, P., Lewis, S. E., & Mungall, C. J. (2019). Gene Ontology Causal Activity Modeling (GO-CAM) moves beyond GO annotations to structured descriptions of biological functions and systems. *Nature Genetics*, *51*(10), 1429–1433. https://doi.org/10.1038/s41588-019-0500-1

Uschold, M. (2018). *Demystifying OWL for the enterprise*. Springer.

van Slyke, C. E., Bradford, Y. M., Westerfield, M., & Haendel, M. A. (2014). The zebrafish anatomy and stage ontologies: Representing the anatomy and development of Danio rerio. *Journal of Biomedical Semantics*, *5*(1), 1–11. https://doi.org/10.1186/2041-1480-5-12/FIGURES/4

Vastrik, I., D'Eustachio, P., Schmidt, E., Joshi-Tope, G., Gopinath, G., Croft, D., de Bono, B., Gillespie, M., Jassal, B., Lewis, S., Matthews, L., Wu, G., Birney, E., & Stein, L. (2007). Reactome: A knowledge base of biologic pathways and processes. *Genome Biology*, *8*(3), R39. https://doi.org/10.1186/GB-2007-8-3-R39

Wang, L. L., Gennari, J. H., & Abernethy, N. F. (2016). An Analysis of Differences in Biological Pathway Resources. *International Conference on Biological Ontology*, Corvallis, OR.

Wang, R.-S., Saadatpour, A., & Albert, R.´ka. (2012). Boolean modeling in systems biology: An overview of methodology and applications. Physical Biology, *9*. https://doi.org/10.1088/1478-3975/9/5/055001

Wang, V. Y., Nielsen, P. M. F., & Nash, M. P. (2015). Image-based predictive model-ing of heart mechanics. *Annual Review of Biomedical Engineering, 17*, 351–83. https://doi.org/10.1146/annurev-bioeng-071114-040609

Welsh, C., Nickerson, D. P., Rampadarath, A., Neal, M. L., Sauro, H. M., & Gennari, J. H. (2021). libOmexMeta: Enabling semantic annotation of mod-els to support FAIR principles. *Bioinformatics, 37*(24), 4898–4900. https://doi.org/10.1093/bioinformatics/btab445

Westerhof, N., Lankhaar, J. W., & Westerhof, B. E. (2009). The arterial Windkessel. *Medical & Biological Engineering & Computing, 47*(2), 131–141. https://www.ncbi.nlm.nih.gov/entrez/query.fcgi?cmd=Retrieve&db=PubMed&dopt=Citation&list_uids=18543011

Wilkinson, M. D., Dumontier, M., Aalbersberg, Ij . J., Appleton, G., Axton, M., Baak, A., Blomberg, N., Boiten, J. W., da Silva Santos, L. B., Bourne, P. E., Bouwman, J., Brookes, A. J., Clark, T., Crosas, M., Dillo, I., Dumon, O., Edmunds, S., Evelo, C. T., Finkers, R., ... Mons, B. (2016). Comment: The FAIR Guiding Principles for scientific data management and stewardship. *Scientific Data, 3*, 1–9. https://doi.org/10.1038/sdata.2016.18

Yang, J. H., Wright, S. N., Hamblin, M., McCloskey, D., Alcantar, M. A., Schrübbers, L., Lopatkin, A. J., Satish, S., Nili, A., Palsson, B. O., Walker, G. C., & Collins, J. J. (2019). A white-box machine learning approach for revealing antibiotic mechanisms of action. *Cell, 177*(6), 1649–1661.e9. https://doi.org/10.1016/j.cell.2019.04.016

Yu, T., Lloyd, C. M., Nickerson, D. P., Cooling, M. T., Miller, A. K., Garny, A., Terkildsen, J. R., Lawson, J., Britten, R. D., Hunter, P. J., & Nielsen, P. M. F. (2011). The physiome model repository 2. *Bioinformatics (Oxford, England), 27*(5), 743–744. https://doi.org/10.1093/bioinformatics/btq723

Zaman, M., Faber, C. L., & Schwartz, M. W. (2022). Homeostasis: A therapeutic target for type 2 diabetes? Annual Review of Pharmacology and Toxicology, *62*, 55–84. https://doi.org/10.1146/annurev-pharmtox-052220-010446

Index

Note: **Bold** page numbers refer to tables and *italic* page numbers refer to figures.